The Ends Game

Management on the Cutting Edge series

Edited by Paul Michelman

Published in cooperation with *MIT Sloan Management Review*

The AI Advantage: How to Put the Artificial Intelligence Revolution to Work
Thomas H. Davenport

The Technology Fallacy: How People are the Real Key to Digital Transformation
Gerald C. Kane, Anh Nguyen Phillips, Jonathan Copulsky, and Garth Andrus

Designed for Digital: How to Architect Your Business for Sustained Success
Jeanne W. Ross, Cynthia Beath, and Martin Mocker

See Sooner, Act Faster: How Vigilant Leaders Thrive in an Era of Digital Turbulence
George S. Day and Paul J. H. Schoemaker

The Leader in a Digital World: From Productivity and Process to Creativity and Collaboration
Amit S. Mukherjee

The Ends Game: How Smart Companies Stop Selling Products and Start Delivering Value
Marco Bertini and Oded Koenigsberg

The Ends Game

How Smart Companies Stop Selling Products
and Start Delivering Value

Marco Bertini and Oded Koenigsberg

The MIT Press
Cambridge, Massachusetts
London, England

This book was set in Stone Serif and Stone Sans by Westchester Publishing Services. Printed and bound in the United States of America.

Library of Congress Cataloging-in-Publication Data

Names: Bertini, Marco, author. | Koenigsberg, Oded, 1964– author.
Title: The ends game : how smart companies stop selling products
    and start delivering value / Marco Bertini and Oded Koenigsberg.
    Description: Cambridge, Massachusetts : The MIT Press, [2020] |
    Series: Management on the cutting edge | Includes bibliographical
    references and index.
Identifiers: LCCN 2019058447 | ISBN 9780262044349 (hardcover)
Subjects: LCSH: Customer relations--Management--Forecasting. |
    Customer services--Forecasting. | Consumer goods--Forecasting.
Classification: LCC HF5415.5 .B4837 2020 | DDC 658.8/12--dc23
LC record available at https://lccn.loc.gov/2019058447

10 9 8 7 6 5 4 3 2 1

To my mother Franca and father Roberto, my partner in crime Ester, and three children Àlex, Noa, and Laila.

To my mother Cipora and father Menachem, my partner Tal, and three children Tamara, Amalia, and Ivri.

… It turns out that everything does happen for a reason.
—Marco and Oded

Contents

Series Foreword

The world does not lack for management ideas. Thousands of researchers, practitioners, and other experts produce tens of thousands of articles, books, papers, posts, and podcasts each year. But only a scant few promise to truly move the needle on practice, and fewer still dare to reach into the future of what management will become. It is this rare breed of idea—meaningful to practice, grounded in evidence, and *built for the future*—that we seek to present in this series.

Paul Michelman
Editor in chief
MIT Sloan Management Review

Preface

Most modern organizations strive to take good care of their customers. They deploy market research to gather insights that can help them develop meaningful products and services that stand out from those of competitors. They also think about their customers' journeys to a purchase, trying to engineer experiences that are at once practical and engaging. Finally, today's organizations are constantly tweaking their internal structures and incentives systems to relate and respond better to the market.

All this effort is excellent. Yet, from where we stand, it is only half the battle.

While many businesses tout their innovative spirit and proudly claim to put customers at the heart of what they do, we seldom see the same determination and steadfast focus in the way they convert market potential into a financial return. If given the chance, customers would gladly pay for the solutions to their needs and wants, the "ends" they seek, rather than the means to achieve them. After all, they only buy products and services because they need or want something from them. Yet a combination of neglect, inertia, fear of change, and comfort with the status quo implies that most companies today still promise results but earn revenue on what comes off the factory floor, so to speak. In fact, in our experience, companies seldom challenge *what* they ask their customer to pay for (the revenue model question) and instead obsess over the secondary, far more tactical issue of *how much* customers should pay (the price question). It is as if the rules of the game in a market are inherited and immutable.

We came to write *The Ends Game: How Smart Companies Stop Selling Products and Start Delivering Value* because we are uncomfortable with this picture and want to shift the focus of the discussion. The rules of the game are far from immutable, and they certainly should not be taken for granted. This is increasingly true as modern digital technology delivers sharper insights into what products and services actually do to customers, empowering them to demand accountability. Business leaders and policy makers need a roadmap to develop and implement the right revenue model given this opportunity … and before it becomes a threat.

At the same time, we wrote *The Ends Game* knowing that our student and executive audiences are interested in understanding the motivation for, and difference between, alternative modes of generating revenue. The radical and disruptive recent changes across many industries are rooted in a conscious, technology-enabled shift in revenue models. Accordingly, this interest at times stems from the realization that new technologies can help them "shake things up" in a market, while at other times it stems from an imminent and present threat of disruption by a bold and nimble entrant.

As academics, our first intuition was to turn to the literature for guidance, but here we found excellent treatments of individual "business models" such as subscriptions (the membership economy!) and collaborative consumption (the sharing economy!) rather than a general theory capable of explaining why these models exist and, importantly, how to think about them as a whole and decide what model is best deployed when.

In many ways, *The Ends Game* is about the future of commerce. We argue that accountability is fast becoming the critical currency of competitive advantage. This change is taking place in just about any industry we can think of, albeit at different speeds and to different extents. As such, while all the revenue models that we discuss exist, they are not equally represented. For example, while subscriptions are all the rage at the moment, in our view they are a stepping stone toward other, more customer-focused revenue models rather than the final point. The same applies to collaborative consumption. Accordingly, the book describes a continuum of plausible revenue models that allows organizations to locate their position in the broader scheme of things and identify the way forward. While we are confident about the direction of change, the pace of that change is less clear.

The ideas that we present in *The Ends Game* are shaped as much by our shared perspective as they are by our individual backgrounds and training. We care about managerial decision making, and therefore this book clearly targets organizations: we want business leaders to understand the shifting landscape of competition as well as the threats and challenges that result, and prepare themselves to take a leading role. Yet we are marketers, and our discipline teaches that the root cause of growth, disruption, and decline in any market is the customer. As such, we build our theory from the perspective of customers, clarifying that, if they "suffer" excessively under a given revenue model, then so do organizations. Organizations should not be indifferent: a key message is that taking the traditional "I want to limit risk" position is actually risky as markets leverage technology to evolve.

Marco's training is in behavioral economics, and he relies primarily on experimental methods to derive insights about how managers make pricing decisions or, in turn, how customers respond to the prices they see. Oded's training is in operations, and he develops mathematical models to understand and explain firm actions and incentives. This unorthodox partnership sparked a healthy dialogue, at times more heated than others, singularly focused on understanding where commerce is heading and, of particular importance, why this is the case. In hindsight, the personal growth that resulted from the learning journey we embarked on was probably reason enough to write these pages, and one that we look back on fondly.

Acknowledgments

There are many individuals and institutions that helped us at different moments in time to bring *The Ends Game* to fruition. We are indebted to all of them, and hopefully we do not leave anyone out unintentionally.

First and foremost, we want to thank Frank Luby. We met Frank relatively early in the process, when we knew the general direction that we wanted to take but the ideas and messages were still raw. Frank helped us—if not coached us—through this initial phase, serving as the best sparring partner any author team could ask for. We drew on Frank's experience as a consultant, journalist, and researcher at the onset and virtually across the entire project ... including this acknowledgments section! His guidance was spot on, and this book is in many ways shaped by his contribution.

Next, we are grateful to our home institutions, ESADE Business School and London Business School, for encouraging us and giving us the freedom to think big thoughts. This freedom cannot be understated, as in many educational institutions this is probably not the case—or at least not to the same extent. In addition, we are extremely thankful to London Business School for direct financial support.

Our excellent former students and participants in executive education programs at our respective schools and across other parts of the world played an important role. Unbeknownst to them, they were often our guinea pigs, as we routinely brought new or developing ideas to the classroom for "testing." More important, the passionate discussions with them on revenue models, technology, and everything in between helped us refine our arguments and examples. There is nothing like a room full of inquisitive, impatient minds to keep you on your toes. We thank our varied audiences for pushing us to think hard about the topic and give every angle or twist due consideration.

We also thank our many coauthors and colleagues for adding their individual perspectives: Eyal Biyalogorsky, Simona Botti, Bart de Langhe, Preyas Desai, Kristin Diehl, Sunil Gupta, Daniel Halbheer, Bruce Hardie, Rajeev Kohli, Anja Lambrecht, Natalie Mizik, Elie Ofek, Debu Purohit, Nader Tavassoli, Naufel Vilcassim, and Luc Wathieu. We recall countless discussions on topics that are relevant to what you read in this book and certainly shaped our thinking. Moreover, we want acknowledge the input from all the scholars that each year join us at the Pricing Symposium, a conference that we organize together with Martin Spann from

Ludwig-Maximilians-Universität in Munich. Parts of the book were first presented at the inaugural meeting.

Just as many people in the education sector helped us to fine-tune the concepts that underlie *The Ends Game*, amazing professionals such as Mark Billige, Marieke Flament, Carlo Gagliardi, Yaron Kopel, David Lancefield, Robert Maguire, Amadeus Petzke, and Todd Snelgrove ensured that we kept our feet well and truly on the ground, insisting that our ideas match and serve reality. Their "touch" is particularly evident in part III of the book where we outline the actions that are critical for organizations to succeed in markets increasingly focused on accountability.

We are indebted to Danny Stern, Ania Trzepizur, and everyone else at Stern Strategy Group, who advised us on the processes of developing our ideas into a book and finding a like-minded publishing house that would support and complement us. The world of publishing was new to us, and hopefully this will be the first of many fruitful collaborations.

On that note, we want to thank Emily Taber at the MIT Press for taking a chance on *The Ends Game*. She understood what we wanted to say, and why it matters for organizations, from the onset. Equally important, we also want to thank her for the unwavering support. Emily's feedback on drafts of the book was incredibly fast, constructive, and, importantly, right on the money. She was a pleasure to work with, and we look forward to a repeat performance.

Last, but clearly not least, we are grateful to our respective families. They have been only too patient with us, seamlessly shifting from role to role as supporter, contributor, or hardened critic as the need arose. The journey from idea to the book you now hold may have taken a little longer than anticipated, but hopefully it was worth the wait!

Introduction: From Promises to Proof

Would you rather pay for health care or better health? Would you rather pay for school or education? Groceries or nutrition? A car or transportation? A theater act or entertainment?

Paying by the pill, semester, food item, vehicle, show, and so on is a poor reflection of the value that individual and business customers actually derive from their purchases. Nonetheless, the idea that a company could be compensated for the quality of the outcomes it delivers, rather than the products and services it brings to market, would have been dismissed until recently as utopian academic theory. Reality called for a compromise, one that most organizations have practiced pragmatically for decades: make a living by selling the "means" to customers, and promise that the "ends" they desire will follow.

Recent technological advances are rewriting the rules of commerce by calling this compromise into question. Mobile communication, cloud computing, the Internet of Things, advanced analytics, and microtransactions are making the exchanges between organizations and customers more transparent. By now, most companies have the ability to record consumption events. In some instances, companies can also observe the value customers derive from their purchases rather than infer it at some aggregate level. These developments have empowered customers who struggle to understand what their money buys them to demand accountability rather than accept simple promises. Customers are no longer the passive price takers of yesteryears. Accountability both defines and widens the gap between organizations that can and want to compete on the outcomes that matter to customers, and organizations that are content with perpetuating the status quo.

Some of the firms that earn revenue by selling "means" dismiss this challenge, turning a blind eye and hoping that it is another passing trend. Others scheme to make life even more complicated for customers, making their prices, assortments, and other commercial decisions more ambiguous and thus less comparable across competitors. However, it is hard to justify these approaches as winning plays in the long run. The alternative, of course, is to embrace change and get to work.

The way we see it, accountability is no longer a fashionable marketing slogan. It is a strategic imperative. Progressive firms are collecting and capitalizing on what we call "impact data" in order to better understand when and how customers use their solutions, and how these solutions actually perform. In many sectors, the technology now exists to turn products into seamless services, to record usage occasions, and, significantly, to quantify performance at scale and with precision. These progressive firms are evolving to make commerce far more efficient than it ever was, and in so doing unlocking market potential and positioning themselves to capture the lion's share of the tangible value that materializes.

The Ends Game prepares firms to win in today's increasingly transparent markets. Using in-depth case studies from sectors as diverse and consequential as health care, education, media, automotive, aviation, and mining, we map the relentless evolution of sectors to the point where the money of customers flows to proof rather than promises. The innovative revenue models that we describe here are not a phenomenon at the fringes of the economy. They are redefining entire markets and altering public discourse. *The Ends Game* helps organizations to understand how these shifts affect their futures and how to exploit the resulting opportunities to the fullest. It is a book about the very nature of commerce and the disruption of markets.

Exposing Inefficiency

Commerce starts with customers who seek out organizations that can solve their needs and wants. Individuals may be after a particular sensation, a tangible benefit, or some combination of the two. The same applies to business customers, who typically want to improve their own financial performance, but may also be swayed by less objective considerations.

To appreciate whether an exchange between an organization and its customers is efficient, we first have to understand the necessary conditions for customers to derive value. Figure I.1 below represents this. There are three critical checkpoints, if you will. First, customers have to *access* the solutions that firms bring to market. Clearly, customers cannot derive value if they are blocked, financially or physically, from reaching the products and services that are intended to address their needs and wants. Second, conditional on access, customers have to *consume* these products and services. Again, customers cannot derive value unless they actually experience or make use of the solution offered by a firm. Third, conditional on access and consumption, the product or service has to *perform* as customers expect—that is, it has to solve the need or want satisfactorily.

We claim that an exchange is inefficient when customers experience friction at any one or more of these checkpoints. Traditionally,

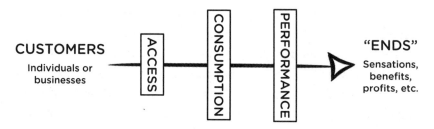

Figure I.1
Customers seek solutions to their needs and wants. Access, consumption, and performance are natural checkpoints toward these solutions.

organizations earn revenue by selling what they make, but this can hinder access, distort consumption, and offers no guarantee of performance. Indeed, a revenue model focused on transferring ownership to customers may appear safe and prudent to the organization, yet it shrinks the opportunity in the market by leaving customers to their own devices. Some customers are priced out of the market or choose to forgo a purchase because it is inconvenient. They may also worry that their tastes will change, and therefore perceive ownership as an unnecessary burden and again decide to stay away. Other customers resolve to pay less to account for the possibility that they will not make sufficient use of the product or service, that it will not perform as advertised, or both.

Fortunately, as shown in figure I.2, firms have options. By lowering barriers to entry and shifting the risk burden away from customers, new revenue models help firms hold themselves to account for access, consumption, and ultimately performance. Specifically, revenue models such as subscriptions, memberships, and "anything-as-a-service" anchor payment to time rather than a physical good or service, opening up the market to profitable customers who may otherwise be out of reach. These "pay-per-time" arrangements address different types of access problems. First, customers may not have the ability to visit a point of sale when the need arises, or they lack the foresight to make an additional purchase before running out of stock. Second, customers may not have the capital to own the product or service outright. Finally, there are a host of categories including music, television, fashion, and books where individual items are inexpensive but variety is important, making the overall purchase a significant one.

Next, revenue models based on unbundling, metering, or sharing all link payment to use, not only in order to expand access, but also to track consumption. In the first case, the firm digitalizes its offering and delivers it in a more granular form. Metering means that the firm supplies the product whole, but charges only for its use. A model built on sharing is one where sellers either manage or join a platform to distribute a product or service across many interested users. Such "collaborative consumption" is growing at an impressive rate.

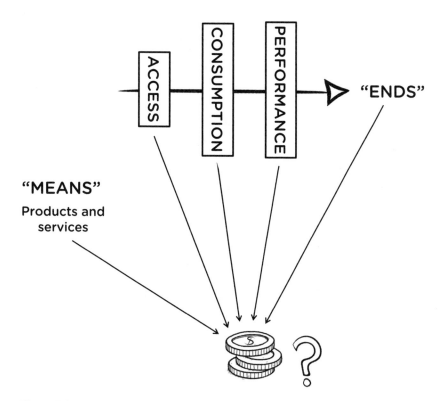

Figure I.2

An organization can earn revenue on what it makes (the "means"), on any one of checkpoints, or on the actual solutions sought by customers (the "ends" themselves).

Finally, revenue models that focus on actual outcomes expand access, mirror consumption, and ensure performance. "Pay-by-outcome" agreements have traditionally found success in contexts where performance is objective, quantifiable, and verifiable. However, in recent years similar agreements have taken hold in consumer markets for health, education, insurance, and even live entertainment.

The Ends Game offers guidelines to business leaders who are willing to meet the challenge of accountability. We provide a framework to understand how innovative arrangements from subscriptions and collaborative consumption to outcome-based agreements align with customers' pursuit of value. We define and clarify the steps to take within and outside the firm, as well as the benefits that accrue from acting.

Clearly, the more an organization aligns the way it earns revenue with the way customers derive value—that is, the more responsibility for the three checkpoints of access, consumption, and performance it takes on—the "leaner" (as in more efficient, less wasteful) the exchange between the two becomes. This is represented in figure I.3, where market potential converts into actual market value as the organization brings its revenue model increasingly into line with the "ends" sought by customers.

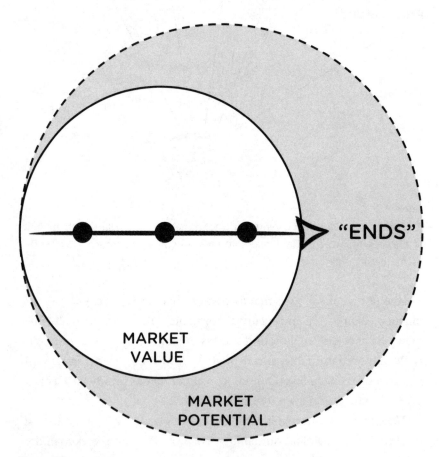

Figure I.3
As the organization moves to align its revenue model with the "ends" sought by customers, efficiency gains convert market potential into actual market value.

The leanest exchange, then, is one where a company's fortunes are contingent on delivering value itself. Several firms that are working closely with customers—such as health care providers, universities, leading industrial manufacturers, and many startups—already stake their positioning and future on this metric. They hold themselves accountable for it. Yet an important lesson in *The Ends Game* is that, while pay-by-outcome may be the final destination, it is not necessarily the *next* destination. Context certainly matters. First, there are obstacles within the firm: what is the definition of "outcome" that we can agree on? How does our offering stack up against those of competitors on this measure? How can we communicate our superiority in a manner that is unambiguous? What is our plan for making the transition from the current revenue model? Second, customers play an integral role: to what extent are they willing to share information about consumption? How motivated are they to play their part in the achievement of a successful outcome? Third, technology will always mark the pace of change.

Accordingly, we do not claim that a shift to, say, outcome-based agreements is urgent and necessary. This overstates the case at the moment. What we do claim, however, is that revenue models anchored on the ownership of a product or service are patently inferior. When selling ownership is the standard in a market, firms with weaker products can siphon off clients with a good story or a cut-rate price. As some firms adopt better metrics, these inferior firms no longer have a place to hide. In the world of lean exchanges between organizations and customers, money flows to proof, not promises. Today, making the transition to a revenue model anchored on time or use is certainly within reach of most businesses as they study and prepare for bolder moves.

What to Expect from *The Ends Game*

Part I provides context for the transition to lean commerce. Chapter 1 traces the evolution of customer focus as a management philosophy and explains why, from our perspective, this evolution is incomplete. From production to distribution to communication, companies have learned to put themselves in the shoes of their customers and reimagine each process with their interests at heart. But customer focus has not reached its logical conclusion. The missing and final part is the process by which organizations convert value to the customer into value for the business.

In chapter 2, we describe how recent technological advancements enable firms to take the logic of customer focus full circle by collecting, analyzing, and interpreting information on when and how customers consume products and services, and how well these offerings actually perform—what we call "impact data." This sets the stage for companies to understand how they can target the various sources of waste—access, consumption, and performance—by choosing a revenue model that is better aligned with how customers derive value from their purchases. Chapter 3 highlights how different revenue models decrease or exacerbate misalignments between how firms create value for customers and how they create value for themselves. Misalignments result in the three interrelated sources of waste that burden so many sectors of our economy and cry out for elimination.

Part II offers a rich set of examples of how companies are already playing the Ends Game, eliminating inefficiencies and making commerce far leaner. Chapter 4 introduces firms that changed their revenue models—or implemented new ones from scratch—to make their products accessible to more customers while minimizing their financial exposure. Chapter 5 shows how firms are unbundling their offerings to make consumption more flexible, tackling access and consumption waste by metering consumption, or activating dormant assets by linking them to the sharing economy. Chapter 6 provides examples of organizations that earn their revenue based on the outcomes they deliver. Firms using these arrangements demonstrate that playing the Ends Game at its highest level changes both the firms' and customers' understanding of what an exchange between them can achieve.

Part III is about action. It focuses on the challenges that firms will face. Chapter 7 offers guidance on how to answer what sounds like a deceptively simple question: *What are we asking customers to pay for?* It provides criteria for defining an outcome that can form the basis for a more efficient revenue model. Chapter 8 explores the reasons why the firms with the greatest chances of success in pursuit of lean commerce are often the most reluctant to play the Ends Game. Several self-imposed obstacles stand in their way.

Chapter 9 explores the issues around impact data, particularly with respect to collection, protection, and use in ways that help customers without endangering their privacy. A company can conceptualize, measure, and charge for an outcome only to the extent that it has access to sufficient, relevant information. Chapter 10 looks at the issues involved in managing customers such that they contribute positively to the quality of an outcome. The elimination of access, consumption, and performance waste often depends on customer motivation and participation, two things that a firm cannot take for granted. It is important to implement mechanisms that ensure customers make positive contributions.

Part III concludes with chapter 11, which examines more of the specific obstacles an established organization faces internally as it starts to play the Ends Game. These include the cultural changes necessary to help the organization overcome its inertia and shift away from tradition. Often it is a newcomer that succeeds in reducing waste by introducing a revenue model conceived to improve access to the market, mirror consumption, or perhaps even guarantee performance. These success stories leave incumbents with a difficult, if not existential challenge that requires an urgent response. If they do not abandon their antiquated practices, they may be forced to abandon their journey as a business entirely.

Prior to the technological breakthroughs of the twenty-first century, lean commerce was not possible on a large scale. There was no practical way to identify and eliminate the billions of dollars of waste that frustrates customers, stifles opportunities, and holds firms back from taking full advantage of their ability to deliver value to customers. Lean commerce is now not only possible, but also a mandate for organizations. The concepts, examples, and suggestions in *The Ends Game* will empower organizations and help them tackle the rocky road ahead.

I   Context

1   Unfinished Business

Think of a retail store that can "market itself to consumers as an 'every-thing' store, with an unrivaled range of products, often sold for minus-cule profit."[1] Fulfilling and sustaining this alluring "everything store" promise would require a relentless focus on customers, striving to sat-isfy everyone's needs and wants. In this particular case, the "company's feel for consumer demand was so uncanny … that it became, for many of its diehard customers, not just the best retail option, but the only one worth considering."[2]

At first glance, this seems to be an accurate description of the mas-sive Chinese retail platform Alibaba, which holds its "Singles Day" on November 11 each year. Alibaba reported that it sold merchandise worth 268.4 billion Yuan (around $38.3 billion) during the 2019 event, spread across more than two hundred thousand brands. Dubbed "the biggest sales event in the history of the planet," Singles Day 2019 easily broke the previous year's record of 213.5 billion Yuan (nearly $30.5 billion).[3] The company's slogan of "Yesterday's best is today's starting line" is a testament to Alibaba's desire to shatter records again and again.

The first paragraph of this chapter could also be an accurate descrip-tion of Amazon, which has more than one hundred million customers in its Amazon Prime program and in 2018 became only the second company in history to achieve a market capitalization in excess of $1 trillion.[4] The numbers from Amazon Prime Day in 2019 show the extent to which Amazon has anchored itself in the minds of customers around the world. Customers purchased over 175 million products in the forty-eight-hour shopping spree, which represents 1,013 products sold per second![5]

Yet the "everything store" we are referring to is not Alibaba. Nor is it Amazon.

It is Sears.

Of course, we mean Sears in its heyday, which seemed to last for most of the twentieth century. We mean the company that began as a mail-order business in the late 1800s and quickly rose to a level of prominence that in its day rivaled what Alibaba and Amazon enjoy today. In fact, this odd juxtaposition of retail giants from radically different eras suggests that companies have repeatedly found success by complementing operational excellence with a relentless focus on customers. This mix has marked the rise of modern commerce since the start of the Industrial Revolution. Thanks to a long list of pioneers, commerce over the last century has changed from an inward-looking activity—making whatever an organization's own skills allowed and then "shoving it" into the hands of customers—to an external one driven by an obsession to understand the attitudes and behaviors of the individuals or businesses that actually purchase.

It is undeniable that, as a management philosophy, customer focus has helped millions of people and enterprises gain unprecedented access to products and services, encouraged consumption by educating markets on an offering's true benefits and range of applications, and fostered the innovation that allows organizations to make increasingly bolder promises of performance, and often meet them. Yet despite all this progress, commerce is still hampered by the type of inefficiencies raised in the introduction of this book.

What is the problem? Why is commerce still so wasteful? In our mind, the issue is not that customer focus is somehow flawed or broken. To the contrary, the issue is that most organizations fail to take the promise of customer focus to its logical conclusion. That is, the typical company today may well obsess over customers when it comes to designing offerings and positioning them in the market. In fact, we struggle to find a business that doesn't praise its customers, and boast of the attention paid to them, on its corporate website. But then the same company pays hardly any attention to customers when it decides how to earn revenue from them. This lapse shrinks the opportunity in the market.

There are at least two reasons why customer focus typically breaks down where it does. First, an organization simply may not appreciate the importance of aligning the revenue model with how customers derive value from their purchases. Accordingly, somewhere after ideation and new product or service development, many companies forget the promises made to customers and fret over internal constraints. What did it cost *us* to bring this solution to market? How much risk are *we* willing to tolerate? What is the desired return on *our* investment?

Second, until recently companies had no meaningful information on how customers interact with, and hopefully derive value from, their purchases. While it is hard to imagine taking a customer-focused approach to the task of generating revenue with no visibility beyond the moment of purchase, it is also true that the importance of gathering information on consumption and performance was largely ignored or dismissed in the past. Again, when the top line is at stake, companies tend to be conservative.

Now, in the twenty-first century, technology offers the opportunity to unravel the mystery, take customer focus full circle, and consequently address much of the waste that persists in commerce. We'll talk more about this game-changing technology and how an organization can harness it in the next chapter. For the time being, however, we want to trace the evolution of customer focus through time. In particular, the next three sections describe the genesis and rise of mass production, mass distribution, and mass communication.

A Car for Every Purse and Purpose

It is impossible to describe and interpret the Industrial Revolution without making substantial use of words such as "mass" and "scale." In the second half of the eighteenth century, Adam Smith became one of the first economists to document and explain the effect of scale on industrialization, linking it to his theories about the division of labor: "The division of labour, however, so far as it can be introduced, occasions, in every art, a proportionable increase in the productive powers of labour."[6] He elaborated on this idea by claiming that the division of labor ultimately leads to the creation of wealth for the masses: "It is the great multiplication of the productions of all the different arts, in consequence of the division of labour, which occasions, in a well-governed society that universal opulence which extends itself to the lowest ranks of the people."[7]

Companies in the nascent industrial world of the 1800s had to learn how to make and sell products economically, so not to price themselves out of the market. For some, the biggest obstacle in fact was figuring out how to manufacture a product in the first place. The bulk of the time, effort, and money went into improving production, both in terms of the physical process on the factory floor and of the mix of features and elements of the product, including engineering, design, quality of the inputs, and consistency. Because competition was minimal, manufacturers felt no pressure to "fit" their product to the unique needs and wants of different customers or customer groups. From their perspective, customers were lucky enough to have an affordable alternative before them.

A perfect example of this is the work of Henry Ford, who introduced the mass-produced car over a century ago. Prior to the introduction of the Ford Model T, cars were clearly a niche product. The bulk of the population could not immediately appreciate the need for them, never mind afford one. Being the visionary that he was, Ford proclaimed, "I will build a motor car for the great multitude. It will be so low in price that no man will be unable to own one."[8] To figure out that low price, Ford did not commission sophisticated market research to understand what his potential customers wanted and were willing to pay handsomely for. Instead, he decided to anchor the price on the costs of production. These costs dropped dramatically as Ford pushed for greater productivity at the assembly line, taking full advantage of Adam Smith's theory on the division of labor and its impact.

Within one year, productivity, measured in terms of the time needed to manufacture a single vehicle, improved from 12.5 hours to only 93 minutes.[9] Accordingly, Ford could afford to lower the price further and crack open the market even wider. Production rose from 10,000 cars in 1908 to 933,720 cars in 1920.[10] And the key to this massive improvement was the resolute consistency of the product. Ford was known to say that customers could own a brand-new Model T in any color they wanted as long as it was black. He had recognized the broad need for affordable, dependable transportation, but refused to indulge in what today we call "personalization" (when the organization adapts the product) or "customization" (when customers do it). The mere existence of the Model T, and its attractive price, were sufficient arguments to drive sales.

Over time, however, production technologies reduced the costs of manufacturing in many markets to the point that the attention of organizations naturally shifted from making "stuff" economically to selling more in the face of rising competition. Customers started having choice, and one way to get their attention, organizations figured, was to get better at giving the public what they actually wanted.

The U.S. automotive industry witnessed this shift in a rather dramatic way in the 1920s. While Ford at the time still wanted to make one car "so low in price that no man will be unable to own one," Alfred P. Sloan, the president and chairman of General Motors (GM), believed the better strategy was to offer one car "for every purse and purpose." This idea led GM to product a tiered range of cars, with Chevrolet at the low end of the quality and price spectrum and Cadillac at the high end. When Sloan took over at GM, Ford boasted a market share of 60 percent, compared to the small 12 percent held by GM.[11] By 1931, however, GM had become the world's largest automaker, a title it held for the next seventy-seven years until Toyota took over.[12]

The combination of operational excellence with the realization that addressing customer tastes can pay off made the difference. The subheading of a 2008 article in *Automotive News* summed up the situation succinctly: "Roaring Twenties made it clear that people were buying status and novelty, not just a ride."[13] That article cited several reasons for the rise of GM, whose Chevrolet brand made vast inroads into the dominance of the Model T, including the development of flexible manufacturing techniques, the growing influence of styling, and the impact of advertising on the relationship of the customer to the product.[14]

Trains, Planes, and Automobiles

Once organizations figured out how to manufacture assortments of goods in a cost-effective, consistent manner, the focus shifted once again. Distribution stepped in as the new bottleneck in business. Indeed, when competition first enters a market, most companies tend to rush to make their offerings stand out in a way that is meaningful to customers. But as companies slowly exhaust the obvious avenues for differentiation, they then work to make their "go to market" approach—the structure by which products and services reach those interested in buying—more pervasive and convenient to customers.

The rise of Sears is representative of the evolution of mass distribution. A *Smithsonian Magazine* article summed up the company's legacy in one sentence: "Sears taught Americans how to shop."[15] For the first few decades of its existence, Sears was the Amazon of its day, making an unprecedented array of goods available to rural America. In fact, just as Amazon originally focused its attention on books, Sears started off in 1893 selling just one product: watches.[16] It also did not have a physical, bricks-and-mortar presence until it opened the first department store in 1925. The company's own statements in the early 1900s demonstrate a strong desire to improve access through scale. "We are able by reason of our enormous output of goods to make contracts with representative manufacturers and importers for such large quantities of merchandise that we can secure the lowest possible prices," Sears claimed.[17] In this case, price was the handmaiden of distribution, with a simple objective (to lower prices relentlessly) achieved by means of massive scale.[18] Sears's dominance as a retailer and catalog house endured into the

1970s, when roughly half of all households in the United States possessed the company's proprietary credit card.[19]

Of course, one could argue that mass distribution predates mass production because people have been shipping bulk agricultural commodities over long distances for centuries. Whether we look at the Silk Road, the Hanseatic League, or the ocean-going European fleets of the Middle Ages, the goal of finding and exploiting more profitable trade routes is nothing new. However, agricultural commodities aside, the true emergence of large-scale, organized distribution has its roots in the new enabling technologies of the late nineteenth century and early twentieth century. One of those great technologies was rail transportation. In the United States, as was the case elsewhere in the world, the railroad played a critical role in helping individuals gain access to the goods they desired in timely fashion. One study from 1944 concluded: "It was far more important that the railroad brought transportation to areas that without it could have had scarcely any commercial existence at all.... Historically, the very existence of most American communities and regions, of particular farms and industrial firms and aggregates, was made possible by the railroad."[20] Marshall McLuhan, the late Canadian philosopher, recognized the transformative role of railroads, writing: "Although America developed a massive service of inland canals and river steamboats, they were not geared to the speeding wheels of the new industrial production. The railroad was needed to cope with mechanized production, as much as to span the great distances of the continent."[21]

The growth of the city of Chicago is a microcosm of the converge of mass production and mass distribution. The Pulitzer Prize-winning poet Carl Sandburg began and ended his poem "Chicago" by describing the city as "Hog Butcher for the World, Tool Maker, Stacker of Wheat, Player with Railroads and the Nation's Freight Handler."[22] Not coincidentally, Chicago was the headquarters of direct-mail retail giants such as Spiegel (founded in 1865), Montgomery Ward (1872), and of course Sears, as well as a leading manufacturing and printing center. And while its share of rail freight has ebbed and flowed over the last 150 years, Chicago continues to enjoy a dominant position: in 2018, half of all

U.S. rail freight shipments passed through the city.[23] Trucking has long since eclipsed rail as the leading mode of freight transportation in the United States, but the gap is surprisingly small: in 2017, trucking represented 40 percent of all shipments, as measured in ton-miles, while rail accounted for just under 33 percent.[24]

Competing for Eyeballs

With production and distribution seemingly "resolved," organizations turned their attention to yet another preoccupation. Instead of asking "How do we manufacture and deliver the products customers truly want in an economical way?," they started asking "How can we best compete for customer attention?"

Organizations realized that they needed a more nuanced commercial recipe in order to break through the clutter of competition and nurture demand in the market. This recipe would complement the efforts already spent on improving and tailoring products and services according to customer needs and wants by fine tuning the more "intangible," brand-related elements of an offering. The objective was clear: to position one's products and services proactively in the minds of existing and potential customers. Accomplishing this goal was important enough, and potentially lucrative enough, that companies began to engage in structured market research and seek greater sophistication in the way they communicated. Mass advertising was born.

One can argue that the United States in the late nineteenth century was fertile ground for professional communication and advertising to take hold. Stressing the role of books as the first modern mass-produced media product, McLuhan noted, "The homogenizing power of the literate process had gone further in America by 1800 than anywhere in Europe. From the first, America took to heart the print technology for its educational, industrial, and political life."[25] The emergence and subsequent adoption of other technologies such as the phonograph, film, and radio "had far-reaching effects upon American society. They broke down the isolation of local neighborhoods and communities and ensured that for the first time all Americans—regardless of their class, ethnicity, or locality—began to share standardized information and entertainment."[26]

Mass advertising—and the modern marketing function as we know it—arose concurrently with these technologies. Packaged-goods manufacturers such as Kimberly-Clark and Palmolive, and other familiar brands including Goodyear Tires, grew swiftly in the early part of the twentieth century thanks to national advertising campaigns. Moreover, these campaigns often prompted the agencies of record to conduct further testing to document what today we refer to as "use cases." Some use cases had a remarkable impact on a business. For example, the original purpose of Kleenex, which was launched in 1924 by a unit of Kimberly-Clark, was as a disposable remover for cold cream and makeup. In 1930, the advertising agency Lord & Thomas recommended that Kimberly-Clark conduct research to understand how people actually use the product in their everyday lives. Much to the company's surprise, the exercise revealed that customers used Kleenex primarily to wipe their noses, not to remove makeup. This basic insight led to the repositioning of the product as a disposable handkerchief, and within a year it became a major contributor to the bottom line of Kimberly-Clark.[27]

Within a relatively short time, the use of market research to better understand customers and create communication material that resonated with them became more popular and more "scientific." According to Claude Hopkins, arguably one of the great pioneers of advertising, "guesswork is very expensive."[28] In fact, one reason for the interest in rigor was the direct experience of agencies and copywriters on work commissioned by direct-mail companies. Hopkins described this experience with a nod toward the same obsession over operational excellence that propelled the birth of mass production and mass distribution: "The severest test of an advertising man is in selling goods by mail. False theories melt away like snowflakes in the sun. The advertising is profitable or it is not, clearly on the face of returns. Figures that do not lie tell one at once the merits of an ad. This puts men on their mettle. All guesswork is eliminated. Every mistake is conspicuous.

There, one learns that advertising must be done on a scientific basis to have any fair chance of success. And he learns that every wasted dollar adds to the cost of results."[29]

One significant step in the pursuit of rigor came in the 1930s thanks to Dr. George Gallup. Born in Iowa in 1901, Gallup earned a Ph.D. from the University of Iowa and was teaching journalism classes when he began to research public opinion extensively. His early work, which included surveys of hundreds of ad executives, revealed that advertisements stressing a product's efficacy and low price ranked high with the executives but carried "much less weight with the public." He used these findings to advocate a more empirical approach to understanding customer preferences.[30] He joined the advertising agency Young & Rubicam in 1932. Today, the eponymous firm Gallup later founded claims to know "more about 'the will' of 7 billion employees, customers, students and citizens than any other organization."[31]

Sears was no outsider to this trend, as the company learned to combine its own operational prowess with a deeper knowledge of its customer base. In fact, the company's success caught the attention of Peter Drucker, the legendary Austrian-born management thinker, who used it as one of his positive examples of how large organizations can learn to turn their attention away from itself and onto the markets they intend to serve. "True marketing starts out the way Sears starts out—with the customer, his demographics, his realities, his needs, his values," Drucker explained. "It does not ask, 'What do we want to sell?' It asks, 'What does the customer want to buy?' It does not say, 'This is what our product or service does.' It says, 'These are the satisfactions the customer looks for, values, and needs.'"[32]

Advertising and marketing executives condensed their own groundbreaking findings into catchy concepts that survive to this day, including the popular notion of a "unique selling proposition." Advertising executive Rosser Reeves came up with the idea in the late 1950s and introduced it in his book *Reality in Advertising*. He used a precise definition.[33] First, "each advertisement must make a proposition to the consumer. Not just words, not just product puffery, not just show-window advertising. Each advertisement must say to each reader: 'Buy this product, and you will get this specific benefit.'" Second, he stressed that the "proposition must be one that the competition either cannot, or does not, offer. It must be unique—either a uniqueness of the brand or a claim not otherwise made in that particular field of advertising." Noting that these two criteria are often easily met, he included the "mass" aspect with his third point: "The proposition must be so strong that it can move the mass millions, i.e. pull over new customers to your product."

Over the last three decades, organizations have intensified their orientation toward customers. Market research techniques continue to make inroads in understanding the true motivations of customers. The processes by which products and services are tailored to fit these individual motivations are also increasingly sophisticated. Amazon claims in its boilerplate that it is "guided by four principles: customer obsession rather than competitor focus, passion for invention, commitment to operational excellence, and long-term thinking."[34] Echoing the early iterations of Sears and Ford, another global retail powerhouse, Sweden's IKEA, operates under the following vision: "To create a better everyday life for the many people. Our business idea is to offer a wide range of well-designed, functional home furnishing products at prices so low that as many people as possible will be able to afford them."[35] However, as we will see next, the pursuit of customer focus as we commonly know it paradoxically stunted a similar evolution in the way organizations convert the value they create for customers into a financial return.

What about the "Back End" of Commerce?

Advertising legend David Ogilvy's famous remark that "any damn fool can put on a deal, but it takes genius, faith and perseverance to create a brand" typifies the perception that, while activities such as branding and communications are inherently strategic, creative, and difficult, the activities that lead to revenue are tactical, formulaic, and require little more than simple arithmetic.[36] The "back end" of commerce, so to speak, is all about running the numbers.

Perhaps the most streamlined execution of a "deal" took place close to a century before Ogilvy's quote, when American merchant and retail pioneer John Wanamaker invented the price tag in 1861.[37] Wanamaker's motivation for the price tag, the label that declares the purchase price of an item for sale, came from his economic sensibilities and attitude toward the standard practice of haggling, which he considered "inefficient and discourteous."[38] Price tags quickly became the norm because they greatly facilitated the ongoing push for customer focus in production, distribution, and communication. Indeed, it is hard to imagine the success of Montgomery Ward or Sears without a fixed price printed in the catalogs they shipped across the United States.

But this invention presented a new challenge. How do you decide on a price for all customers well in advance of an actual transaction? And how do you aggregate the likely differences in the willingness to pay across customers into a single number that is then printed for all to see? These seem like difficult questions to answer under the best of circumstances, let alone under those faced by most organizations throughout the better part of the twentieth century. Accordingly, organizations gradually shifted their pricing decisions away from customers and what they value, which was the focus of haggling, to the one piece of information they could trust and readily collect: information on the cost of making an offering and bringing it to market.

In essence, prices became a mere bystander in a company's efforts to build closer ties with customers. Rather than considering how the process of generating revenue could assist in this endeavor, the company concerned itself with how to cover its costs and minimize any interference. The price tag went along for the ride, treated as a tactical afterthought, but this in turn cemented the idea that a company's only logical move is to earn revenue from selling the "stuff" it makes.

In the next chapter, we look at how breakthrough technologies are drastically changing this perception. The back end of commerce is an organization's unfinished business, and these developments serve as the catalyst for eliminating the access, consumption, and performance waste that lingers in today's marketplaces.

2 Beyond Needs and Journeys

The successful development of a new medication can take ten years or more, with much of that time consumed with clinical trials.[1] But despite all that due diligence, medications raise a whole new set of questions when they finally hit the market: how many patients fail to take their medication properly, or even take it at all? Did the intended patient take the pill or did someone else? How many patients split pills? To make those determinations, should a doctor accept a patient's word, or look at how often prescriptions get refilled and then draw conclusions? Or should the doctor and the health-care system ignore those questions as long as the patient shows therapeutic improvement?

These unanswered questions have already led to the development of a breakthrough pharmaceutical product: Abilify MyCite. This drug, approved by the U.S. Food and Drug Administration (FDA) in 2017, is a pill containing an ingestible sensor that digitally tracks whether patients have taken their medication.[2] An FDA press release explains that the MyCite system "works by sending a message from the pill's sensor to a wearable patch. The patch transmits the information to a mobile application so that patients can track the ingestion of the medication on their smart phone. Patients can also permit their caregivers and physician to access the information through a web-based portal."[3] What the Abilify pill makes available is what we call *impact data*.

Impact data shed light on when and how customers consume products and services, and how well these offerings actually perform. Impact data are the missing link in understanding the value a customer ultimately derives from a given purchase. Customer focus, as managers commonly understand it, emerged from the quest by businesses to identify their customers' needs and wants. Over the last decade, companies have made progress in mapping these motivations as well as customers' purchase processes (the so-called decision journey) and experiences. However, prior to the widespread availability of twenty-first-century information technologies, a company could not observe post-purchase behavior directly, completely, and in real time.

Now companies need to collect and use impact data to close the loop on customer focus. Every company has similar sets of questions that are impossible to address decisively without impact data: How often and how well do our customers use our solution? Where and why do they use it? How satisfied are customers?

Impact data add transparency, allowing organizations to pinpoint changes in behavioral patterns and draw more reliable conclusions about why they are happening. Organizations can improve their products and services to generate more value for themselves and their customers. The collection and interpretation of impact data create vast opportunities to improve exchanges, because the resulting insights allow an organization to adopt a revenue model that is more efficient.

At the same time, impact data create a brand new set of obligations for an organization, because they are extensive, personal, and may even expose behaviors that customers prefer to keep private. Will the company use impact data to exploit the customer relationship? Or will it use them to make commerce more efficient? These obligations create a mutual demand for accountability. Customers can demand that companies use impact data in their individual interests. Knowing they can make in-depth comparisons more easily than ever before, customers will naturally gravitate to sellers that adopt a revenue model best aligned with the value they derive. To paraphrase the old saying, customers can finally get precisely what they pay for, no more and no less. Companies in turn can demand accountability from customers to ensure that they use the product or service in a way that achieves the best outcome. Finally, and most importantly, companies can (and should) demand accountability from themselves, making a commitment to leverage impact data only to help their exchanges with customers.

Why Impact Data Matter

Think about the car odometer. "How many miles does it have on it?" is one of the first questions a mechanic will ask when someone brings in a vehicle for service. It is also one of the most important questions a potential used-car buyer will ask. It is obvious that all miles driven are not created equal. What is not obvious, and what until recently was impossible to understand, is how different each mile driven actually is. An aggregate number of miles does not reveal who had access to the car, the conditions under which those miles were driven, and how well the vehicle performed for each individual mile. An aggregate number of miles also offers no insights into miles *not* driven because the car had some issue that made it temporarily inaccessible.

Nonetheless, both the mechanic and the potential buyer are likely to endow a car's odometer reading with many layers of meaning. That single number sets expectations on wear and tear, repair needs, residual value, and the presumed intensity of usage relative to the car's age. Car manufacturers even build assumptions about mileage data over time into their warranty offers. For example, in the United Kingdom the Volkswagen Golf, one of the best-selling vehicles in the world, currently qualifies for a three-year warranty consisting of a two-year unlimited mileage warranty and a third-year warranty with a 60,000-mile limitation.[4]

Now imagine that cars do not have odometers. How difficult would it be to make the same judgments about car usage and conditions, never mind draw useful conclusions from them? In lieu of a mileage reading, one could ask the current owner for impressions about his or her car usage, which in turn could serve as input to some back-of-the-envelope mathematics. One could examine the car for obvious signs of intensive or prolonged use, such as rust, worn parts, or unusual sounds. But no matter how much intuition and guesswork one applies, the truth is that no one could know anything about the car's usage or performance with certainty. Once the vehicle left the dealer's lot for the first time, its true ongoing history would be a mystery.

What solves the mystery is impact data. The individual customer in possession of the product may have some insight into how the product was used, but only impact data—collected independently of the customer's memory and beliefs—can provide a complete, precise, and objective understanding. In fact, the absence of impact data can trigger vast inefficiencies: how many consumption opportunities are forgone because someone needed a product or service but could not access it? Beyond that, how many goods are used far beyond their intended or useful life spans? How many individuals or businesses fail to fully amortize their monthly or yearly subscriptions? How many customers buy items they rarely or never use? And finally, how many consumers invest in products or services that do not create the outcomes they were supposed to?

Impact data replace anecdote and guesswork. As we described in chapter 1, organizations have numerous proven means at their disposal to decipher customers' needs, wants, and decision journeys. But they could not determine what truly happened beyond the purchase. Who was the ultimate user of the product or service? What did a customer really do after gaining access to it? How well did that product or service perform relative to its promised benefits?

Without impact data, companies need to draw inferences from measurements of repeat business, such as renewal rates, and measurements of customer loyalty, such as the Net Promoter Score (NPS), which is based on answers to one single question: "How likely is it that you would recommend [company X] to a friend or colleague?" The father of the NPS, Bain partner Fred Reichheld, explained its rationale in the *Harvard Business Review* in 2003: "By substituting a single question for the complex black box of the typical customer satisfaction survey, companies can actually put consumer survey results to use and focus employees on the task of stimulating growth."[5] Such proxies may have served their purpose well, but companies no longer need to make assumptions and draw inferences when data on consumption and performance are readily available.

Where Impact Data Come From

To tell the full story of how technology creates this newfound transparency, we need to understand the technological advancements of the last two decades, and especially the last five years. These advancements have fundamentally altered the flows of information between organizations and their customers and, at the same time, altered the very nature of products and services.

These technologies fall into three broad categories: hardware, connectivity, and intelligence. The most basic hardware includes the physical objects that collect and transmit the data. Common examples are sensors and scanners, which detect and measure changes in an environment (temperature, moisture, motion, heart rates, speed, pressure, and volume, to name only a few.) Other major hardware components are communications devices that allow the transmission or receipt of data or both. Products that contain such hardware are often referred to as "smart products."

Connectivity refers to the networks that facilitate the exchange of large amounts of data. These include the bandwidth provided by telecommunications networks (3G, 4G, 5G), cable networks, and the access to storage and applications provided by virtual or cloud computing.

Intelligence refers to how someone or something that receives impact data can transform, analyze, interpret, and apply them. It includes not only applications that make data useful to a customer but also more advanced areas such as artificial intelligence (AI) and machine learning. AI in its most basic sense is the use of computers and algorithms to emulate higher functions that humans normally perform, such as facial recognition, pattern matching, and decision making. Machine learning, fundamentally, is the process that enables these algorithms to improve themselves as they process more data.

Fitness enthusiasts show how people can benefit from this combination of hardware, connectivity, and intelligence. The heart-rate monitor strap houses the hardware (sensor and transmitter). During the workout, the smartphone—using a telecom, local area, or WiFi network—picks up the impact data transmitted by the monitor. The smartphone has an application (app) that presents the data in a useful form and in real time to the user. Most apps allow the user to store the data after a workout and send the data to other parties either manually or by opting into real-time transmission.

An AI app in this case could recommend future workouts based on impact data as well as on personal information such as weight changes, diet, fitness objectives, health parameters, and the performance records of the user and others. It could also detect anomalies in performance and generate incentives for improvement. In an analog world, a human being would have needed a stethoscope, a scale, plenty of paper, a very active imagination, and lots of patience to accomplish the tasks that AI performs instantaneously and far more reliably. Machine learning would enable that AI algorithm to improve itself and self-correct as it receives more data and better understands the relationships between inputs and outputs.

The Internet of Things (IoT) refers to the communication conducted by and between nonhuman devices. The technology consulting firm Gartner predicts that from 2019 to 2021 the number of connected "things" will almost double to 25 billion.[6] These devices generate huge amounts of data, including where they are located (tracked by GPS technology), how they are performing, what they are experiencing (tracked by a wide variety of sensors on or near the devices), and who is communicating with them (tracked by anything from a mobile connection to voice or facial recognition.)

An example of how these technologies work together is the combination of a modern car and the "intelligent pavement," which consists of concrete slabs "embedded with an array of sensors, processors, and antennae."[7] In 2018 the state of Colorado planned to test this duo on a stretch of highway. Thanks to the sensors and the communication technology, the pavement and all the cars traversing it can "read" each other and "talk" to each other, transferring data on vehicle weight, type of vehicle, speed, road conditions, and many other factors such as information on the driver, passengers, and cargo. All these interactions can be transmitted to and stored in the cloud, where they can be used to identify real-time situations such as traffic density and potential risks. Parties with access to the data can apply algorithms to identify underlying patterns, spotting everything from whether automated road signs need to inform drivers of hazards, impending weather, or sudden changes in traffic conditions to whether an individual driver may be impaired or engaged in dangerous behavior such as weaving, driving too fast or too slow, or driving on improperly inflated tires (inferred either from sensors in the tires or the weight distribution of the vehicle on the pavement).

These technologies are so pervasive, and the data they collect and transmit so extensive, that the preceding example could apply to the interaction of any device with any individual provided the requisite hardware, connectivity, and intelligence are present and engaged. Impact data not only drive transparency between organizations and customers, but also democratize it. The challenge to companies is whether they will use that better understanding of consumption and performance to make commerce more efficient. This is a question of accountability.

How Impact Data Change the Rules of the Game

How organizations employ this newfound power is the essence of the Ends Game. For the first time ever, technology makes it possible to measure and understand consumption, product and service performance, and their interrelated patterns in real time, at a detailed personal level, and at scale. The technologies described above enable the tracking, storage, and analysis of immense amounts of impact data, which ultimately enables companies to bring their revenue model in closer alignment with the way customers derive value from their purchases. Companies are holding themselves accountable when they shift their emphasis away from selling ownership to charging for access, consumption, and possibly performance itself.

Think back to the odometer example. A mileage reading of, say, 55,000 miles says nothing about the patterns of consumption that led to this point, how they have changed over time, and the outcomes that driving enabled. How many trips were long vs. short in terms of distance and time? What were their purposes? Then there is the question of which individuals drove the vehicle, and when. Every driver has idiosyncrasies that can affect a vehicle's wear and tear and performance. Tesla has recognized that fact, and now offers drivers the ability to store their personal settings or "driver profile" in the cloud, then download it into any Tesla vehicle they may drive.[8] The car will automatically change the seat, suspension, mirror positions, and other features. Finally, the rich contextual detail matters. A six-hour trip of 300 miles in the winter, with four passengers and a full load of luggage, over rough, hilly, winding roads will have a vastly different effect on a vehicle than a similar trip on a perfect summer day with one passenger, better roads, and moderate traffic. All miles consumed are not created equal.

Think of the music industry as another straightforward example of how much customer behavior a company can now track, and how recent this phenomenon is. Years ago, when most people purchased compact discs, no one except perhaps the buyer knew what happened to the product once it left the music store. Even then, we think it is a safe bet that absolutely no one can say with accuracy how often he or she has played the individual songs on any given CD, nor where or when. Record labels had no way to tell what customers did with a CD, once they bought it. They didn't even know if the shrink wrap was removed. The customers' listening behaviors were a complete mystery. Nor could anyone know how much people enjoyed a given song. Artists and record labels could conduct surveys or trust anecdotal evidence, but they had no rich, dynamic, large-scale factual basis for making decisions. They had no impact data whatsoever.

Compare that to the knowledge that a streaming service such as Spotify currently collects. In late 2016, it launched an advertising campaign with the tagline "Thanks, 2016. It's been weird" featuring provocative comments such as "Dear person who played 'Sorry' 42 times on Valentine's Day: What did you do?"[9] Spotify collects a large amount of personal data on its users—including information on the nature of their streams (what, where, when, and how)—and stores it on the Google Cloud Platform. Spotify's proprietary algorithms then allow it to infer the "why." As Spotify describes it: "Our system for predicting user music preferences and selecting music tailored to our users' individual music tastes is based on advanced data analytics systems and our proprietary algorithms. The effectiveness of our ability to predict user music preferences and select music tailored to our users' individual music tastes depends in part on our ability to gather and effectively analyze large amounts of user data."[10]

That description may sound self-serving on Spotify's part, but it encapsulates what a company can do when it has access to impact data. By observing actual usage patterns, a company can understand better than ever before why a customer makes certain choices, and therefore determine what needs and wants the customer is trying to fulfill and what outcomes they are trying to achieve. In the case of a music streaming service, a company can use impact data to get a step closer to understanding the value of a song choice—at a given time, place, and context—to an individual listener. That would have been a purely hypothetical exercise in the eras of vinyl, cassettes, or CDs. Real-time information and communications technology offer businesses unprecedented insights at a very detailed level into access, consumption, and outcomes from the customers' perspective. Thanks to these ongoing advancements, some of which are still in their infancy, companies can measure, quantify, and communicate those benefits, often in real time at the individual level, and then design and implement a more efficient revenue model.

Impact Data Shrink the Size of Segments

The Ends Game is best played in the singular, not the plural. Individual impact data allow companies to achieve efficiencies by pursuing the elusive "segment of one." This term is not new. But with the emergence of advanced information technologies, the segment of one is no longer a theoretical aspiration—something to be approached but never achieved. It is now tantalizingly close to becoming mainstream reality.

"Marketers must leverage the power of insight-driven personalization and use a predictive or prescriptive approach to understand the needs and desires of customers," CapGemini wrote in a report in 2018.[11] *Forbes* described how that might work: "From now on, retailers won't use the buyer's segmentation where an audience acts similarly. They can create detailed digital customer profiles (DCP) and personal segments for each user. Machine learning-based algorithms (decision trees) show the retailer with a certain probability whether the buyer will perform the desired action at the suggested price or not. This is the future of retail: not to compete for a customer with price wars, but to fight for data to make the buyer's experience as personal and unique as possible."[12]

Commerce is moving toward this singular world. It is no longer about many customers' needs, wants, and actions. Commerce is now about *one* customer's needs, wants, and actions. As consumer research pioneer George Gallup apparently never tired of saying, there are billions of ways to live a life, and each one should be studied.[13] Thanks to today's scalable technology, impact data are finally available to study any individual life in all its rich detail. The challenge lies in accountability, which means cultivating the relationship between organization and individual in a manner that is sustainable and mutually beneficial.

The right revenue model is what sustains that relationship. When a company knows whether and how a customer—be it an individual or a business—uses a product or service, and how these uses relate to actual outcomes, the company has the opportunity to implement meaningful changes. It can easily charge in smaller, more manageable recurring amounts, thus opening up access to customers who otherwise could not find or afford their goods or services. This reduces access waste. Organizations can also cut consumption waste, because they can now track and influence usage in real time. For the first time, they can also conceptualize and measure the actual outcomes that customers derive, which can lead to revenue models that eliminate performance waste.

Impact Data Replace Anecdote and Guesswork

Give the same customer the same product twice, and there is no guarantee that you will see the same level of consumption, or even the same level of product performance from equal consumption. The difference is "context," a much richer term than "occasion" or "event." Context is the richest possible description of all the conditions that can influence access, consumption, and performance, positively or negatively. Such conditions can be external, such as the weather, time of day, and location, and can also be internal, such as a customer's state of mind, motivation, or current level of preparedness.

Companies are now in a position to understand context in real time and respond to customers accordingly. Automotive safety is a good example. The World Health Organization estimates that road accidents kill more than a million people every year and cause some 50 million injuries.[14] When cars can communicate with one another about traffic, road conditions, or other hazards, they can help drivers avoid accidents. At a personal level, drivers can also receive prompts from their own car about whether they are speeding and about other steps they might take to reduce their risk. The Swedish insurance company Folksam offers drivers such as an option, with the "long-term goal to save lives and reduce the number of traffic accidents." Folksam's incentives could reduce a driver's insurance premiums by as much as 20 percent.[15] The U.S. insurance company Progressive offers a similar program using either a separate device or the vehicle's own telematics.

The powerful combination of real-time consumption patterns, personalization, and rich contextual data—all at scale—provides companies with a basis to establish and reinforce trust with their customers, one by one. That is the desirable outcome of today's technological changes. We see no reason why the power of this combination won't grow as new technologies emerge. But the combination is a dangerous cocktail if companies do not use it in an accountable manner.

3 Leaner Commerce

Imagine the frustration that people feel when they see headlines such as "Shocking Truth: 20 Percent of Health-care Expenditures Wasted in United States and Other Nations."[1] Warren Buffett, the chairman and CEO of Berkshire Hathaway, refers to "the ballooning costs of health care" as "a hungry tapeworm on the American economy."[2]

The trillion-dollar health-care industry is the rule, not the exception, when it comes to economic inefficiency. Across all corners of the global economy, inefficiency is an epidemic, imposing mind-boggling costs on customers, companies, taxpayers, and governments. For example, think of the following three data points about the inefficiency that results from car ownership. First, cars remain parked roughly 95 percent of the time. Second, according to the transportation consulting firm Inrix the cost of lost time attributed to rush-hour commuting in the United States in 2017 was $305 billion.[3] This means that cars spend far more time occupying space than they do moving on roads.[4] Third, Massachusetts Institute of Technology (MIT) professor Eran Ben-Joseph estimates that there are 800 million surface parking spaces in the United States, which represents an area larger than the entire state of Connecticut, or roughly half the size of Belgium.[5] In some cities, parking spaces take up one third of all downtown land. It is no wonder, then, that Michael Szell, a former researcher at the MIT Senseable City Lab, refers to urban parking spaces as a "huge space that's basically wasted." In short, the operational and opportunity costs of vehicle ownership impose an enormous burden on society, and the general public is weighing the pros and cons, as the *Wall Street Journal* described extensively in a 2017 feature with the headline "The End of Car Ownership."[6]

Second, the belief that the advertising sector is inefficient has endured ever since John Wanamaker supposedly said, "Half the money I spend on advertising is wasted; the trouble is I don't know which half." Fast forward to 2017, and a statement from Marc Pritchard, Procter & Gamble's chief brand office, confirms Wanamaker's estimate, at least in spirit: "There's ... at least 20 to 30 percent of waste in the media supply chain because of lack of viewability, nontransparent contracts, nontransparent measurement of inputs, fraud, and now even your ads showing up in unsafe places."[7]

Third, think about education. One hypothesis on the sector—which spends $1.15 trillion per year[8] in the United States alone from government and private sources—is that money is like "pixie dust;" that is, the more one spends on education, the better the results. An exposé by National Public Radio raised some critical questions about the definition of "success" and outcomes in education. Does success mean better test scores? Does it mean higher graduation rates, higher incomes after graduation, or other "life outcomes"?[9] In the 2018 book *The Case against Education: Why the Education System Is a Waste of Time and Money*, George Mason University economics professor Bryan Caplan argues in favor of sharp cuts in educational spending to stop a "wasteful rat race" driven by grades and credentials rather than actual skills. Caplan also recommends more emphasis on vocational training, because "practical skills are more socially valuable than teaching students how to outshine their peers."[10]

Drills or Holes?

The primary cause of this waste epidemic is rooted in the very nature of commerce. Business and individual customers seek out organizations that can solve their needs and wants, and in turn organizations want to generate revenue—and hopefully earn a profit—from bringing possible solutions to market. This sets up the exchange between the two parties, but their interests are not necessarily aligned. The way organizations generate revenue typically does not match with the way customers derive satisfactions.

No customer wants to buy goods, gadgets, components, or instruments for the sake of owning them, yet these products are what most manufacturers attach their price tags to. For example, if a customer wants the benefit of clean clothes without leaving home, she generally has to purchase a washing machine and detergent. She cannot purchase cleanliness and convenience directly. Similarly, no customer wants to buy the time a professional spends on a task or project. Customers want to purchase the outcome, be it a complete and accurate tax return or the lasting repair of a leaky sink. Yet professional service firms and tradespeople insist on invoicing based on time, even when the outcome is uncertain or never achieved in full.

Inefficient revenue models are the norm rather than the exception across most sectors. Digging a little deeper, this inefficiency ultimately comes in three forms: access, consumption, and performance. These three forms or sources of waste are sequentially related, as they represent checkpoints for customers to derive value in a transaction: customers want solutions, and solutions yield value only when they can be accessed, when they are consumed, and when they perform as desired.

More precisely, *access waste* is tantamount to saying "Customers can't get it" because of a financial or physical constraint. Financial constraints exist when, for example, a business lacks the necessary capital to purchase a piece of equipment or machinery, or an individual cannot afford a single large expense (a car) or many smaller expenses (a music collection or wardrobe). Physical constraints can exist when reaching or replenishing a product or service is inconvenient (effortful, time consuming, etc.). At the same time, customers may not only have trouble reaching a specific solution, but also getting rid of it. If their preferences change over time, they are stuck, and disposal can be expensive. *Consumption waste*, on the other hand, is tantamount to saying "Customers don't or can't use it." Common forms include situations where customers are required to buy a larger quantity than they actually need, or when a single asset is underutilized. A business or individual might purchase an expensive asset that sits idle the vast majority of the time. It may also be that the customer is ready to use a product or service, but at that specific moment it is not working properly. Finally, *performance waste* occurs when the product or service doesn't deliver the value expected from it. The customer has access to it and consumes it, but the end result simply isn't satisfactory.

Marketing guru Theodore Levitt hinted at this misalignment between organizations and customers when he famously observed that people "don't want to buy a quarter-inch drill, they want a quarter-inch hole."[11] A revenue model that emphasizes the transfer of ownership of a product or service from the organization to customers presumes that there is a direct, strong link between the amount spent on a purchase (buying the drill) and achieving the desired outcome (the hole in the wall). This is evident also in the public sector. It drives the thinking that higher teacher salaries and better facilities improve education, that greater investment in police makes a city safer, that more spending on medicine makes a society healthier, and that investment in infrastructure makes a city or region more attractive and economically viable. However, greater spending does not necessarily lead to greater satisfaction, directly or even indirectly. Ownership in and of itself does not guarantee access, nor does it guarantee consumption. It certainly does not guarantee performance. In fact, ownership may represent a hindrance to customers, deterring some from entering the market altogether and exposing others to unnecessary risk.

In an oft-cited passage from the mid-1960s, Peter Drucker described this issue and the awkward compromises made to resolve it: "The customer rarely buys what the business thinks it sells." The rest of that passage, however, is far more revealing and important to understand how selling ownership can be inefficient: "One reason for this is, of course, that nobody pays for a 'product.' What is paid for is satisfactions. But nobody can make or supply satisfactions as such—at best, only the means to attaining them can be sold and delivered."[12]

Customers thought they were paying for satisfactions (clean clothes, holes in a wall, etc.), but in reality they were paying for washing machines, drills, and the *promise* that they would work. Buyers want satisfactions, but sellers can neither deliver it directly, nor charge for it directly.

Lean Commerce: The Momentum toward Better Metrics

The time has come for organizations to accept the idea that much of the inefficiencies we observe across many parts of the economy is the direct result of misaligned incentives. It follows, then, that striving to narrow the gap between the way organizations earn revenue and the way customers derive value can mitigate waste, or even eliminate it. That is what we mean when we talk about leaner commerce. The greater the alignment between organizations and customers, the leaner (more efficient) commerce becomes.

Thanks to the technological breakthroughs we described in chapter 2, firms now have a spectrum of revenue models to choose from in order to drive this change. Firms can do much better than simply sell ownership. Similarly, customers can do much better than simply play along with what organizations have to offer and hope for the best. Specifically, one step removed from ownership models are what we call *access models*, which tackle access waste, financial or physical, by breaking down a purchase that may be onerous or hard to complete into a periodic expense, and ensuring that the product or service is available to customers for the duration of the relationship. The increasing popular options of subscriptions and memberships clearly fall into this category. We discuss access models in greater detail in chapter 4.

Next, *consumption models* tackle both access and consumption waste by allowing the customers to pay only when they use the product or service. Flexible consumption comes from unbundling the product or service, metering an individual customer's usage, or allowing owners to share otherwise underutilized assets such as cars, clothes, or dwellings with others. We focus on these three options in chapter 5. Finally, *performance models* tackle access, consumption, and performance waste because the customer pays based on the extent to which an outcome was achieved. The ultimate outcome, of course, is value. Commerce is at its leanest when the metric the company uses to earn revenue is the actual satisfactions that customers derive from their purchases. The conception and implementation of these models, however, present several challenges. We show several successful examples in chapter 6, then elaborate on the challenges in part III of the book.

The Upside of Leaner Commerce

When firms choose a revenue model that corrects the misalignments we have described, they attract more customers by lowering barriers to purchase and boost willingness to pay by progressively taking on the risk inherent in the exchange. They can also get more benefits from their investments in innovation and differentiation, because the value they create is more closely aligned with what the customer pays for. Customers who pay for "ends" (or a good proxy for them) instead of "means" are less susceptible to the risks of lower-cost, lower-quality products. They spend their money on proof, not on promises. These revenue models deliver these benefits, however, only when firms hold themselves accountable to them. Achieving and maintaining that level of accountability is a long-term journey, not a quick fix.

The health-care industry has begun to take the right steps. If 20 percent of health-care spending is indeed wasted, the potential benefits from reducing that inefficiency are staggering. The Organization for Economic Co-operation and Development, which sponsored the underlying study that came to that conclusion, argues that recognizing the extent of this waste is the first step in getting all stakeholders involved to change their behavior.[13] An article published in the *Harvard Business Review* was straightforward about what must happen next: "What leads to cost savings is reorganizing care around the delivery of *health* rather than *health care*."[14] In other words, the ultimate resolution for this immense inefficiency in health-care systems is a focus on outcomes rather than the means of achieving them.

Efforts to focus on outcomes will not only benefit society by reducing waste, but also offer a huge upside for companies that make that journey. In the health-care sector, "Winners ... will be those that build a sustainable competitive advantage through better access to, and analysis of, clinical data [and] through deeper insight about how to improve outcomes," the Boston Consulting Group concludes.[15] In the same report, the Boston Consulting Group describes the consequences for suppliers that fail to focus on outcomes: "Hospitals, drug companies, and device makers that cannot demonstrate that their procedures, medications, and products genuinely add value will suffer."[16]

The companies and institutions investing in these solutions will not necessarily come from within the existing set of suppliers. In January 2018, Amazon, JP Morgan, and Berkshire Hathaway announced an ambitious long-term effort to address the cost of health care. Both Warren Buffett and Amazon CEO Jeff Bezos used the word "outcomes" in summarizing the challenge they have taken on. "We share the belief that putting our collective resources behind the country's best talent can, in time, check the rise in health costs while concurrently enhancing patient satisfaction and outcomes,"[17] Buffett said. Bezos noted that "hard as it might be, reducing health care's burden on the economy while improving outcomes for employees and their families would be worth the effort."

The pursuit of leaner commerce has begun. It will forever change the nature of markets and society at large. Firms that build their revenue models around outcomes—and hold themselves accountable to them— are poised to reduce or eliminate billions if not trillions of dollars in waste from economies around the world, with social and ecological consequences we can only begin to imagine.

II Models

4   Shaving, Rocking Out, and Looking Fabulous

Warren Buffett couldn't hide his excitement when Procter & Gamble announced it would acquire the shaving and personal care giant Gillette for $57 billion. "It's a dream deal," he proclaimed on that day back in January 2005. "[It will] create the greatest consumer products company in the world."[1]

At the time, it was easy to appreciate Buffett's enthusiasm. Five words captured the essence of Gillette's deceptively simple model for its razor and blades business: "add features and raise prices."[2] That model generated a "seemingly indestructible, high-margin revenue stream" with an estimated global market share of 50 percent.[3] One analyst described the potential leverage of the combined businesses over consumers and retailers by saying, "Shelf space is diamond-encrusted gold. It's exposure to the consumer, and everyone wants exposure to the consumer."[4]

The prices for razors and blades, which are complementary goods, have moved along one dimension: the package price. To live up to its slogan of "the best a man can get," Gillette recognized that the only means to reflect the value created through its product innovations was to increase prices along that dimension. A report in the *Financial Times* in July 2013 said that since 1990, the price of Gillette blade cartridges, adjusted for inflation, had risen by 236 percent, or roughly 5.5 percent every year over a twenty-three-year period.[5] But the implementation of that model led to frustration among customers. They had no choice but to "go to the store, ask a salesclerk to open the plexiglass-encased 'razor fortress,' and pay more than seems reasonable for a small pack of blades."[6]

Five years after its acquisition by P&G, Gillette faced what *AdAge* called "an acid test" when it launched its latest innovation, the Fusion ProGlide shaving system.[7] Could Gillette's time-tested "trade-up model" still bring strong financial results in the tough economic times that lingered after the Great Recession in 2008? According to the *AdAge* report, replacement blades for ProGlide would cost 10–15 percent more than the previous model. That price increase was similar to the increase when Gillette upgraded its Mach 3 system.[8]

Little did the *AdAge* reporters, or anyone else in the market, realize that the ultimate acid test for Gillette would have nothing to do with the Fusion ProGlide. Instead, it would have to do with the outcome of a seemingly innocuous discussion at a holiday cocktail party a few months later. One of the party guests, Mark Levine, had a bunch of unsold, surplus razor blades in a warehouse. When another guest, a young entrepreneur named Michael Dubin, heard about that, it reminded him of his ongoing frustrations generated by Gillette's business. The ensuing discussion between Levine and Dubin sparked a question: If a company could mail blades to customers at a reasonable price, would men appreciate the convenience?[9] Both Levine and Dubin thought that men would.

The Dollar Shave Club was born. Instead of paying lots of money for blades locked in the "plexiglass fortress" at a retailer, men who joined the club could pay a few dollars per month to have replacement razor blades shipped to their door on a regular basis. This subscription model's ordering process was nearly effortless, the price a small fraction of what Gillette charged, and the quality "good enough." And it worked.

Thanks to the combination of irreverent and viral advertising, conscious brand building, and unmistakable affordability, Dollar Shave Club (DSC) attracted venture funding almost as quickly as it attracted subscribers. By 2015, the erstwhile startup had amassed more than four million subscribers served by over six hundred employees. With revenue of roughly $150 million in 2015 and a goal of $200 million in its sights for 2016, the company quickly claimed about half of the online market for razor blades.[10] In the meantime, Gillette's share of the overall market for men's razors had fallen for six consecutive years, from more than 70 percent in 2010 to 54 percent in 2016.[11]

One year after DSC hit the market, the entrepreneurs Jeff Raider and Andy Katz-Mayfield launched a rival razor and blades subscription service called Harry's. Their rationale, expressed on the Harry's website to this day, reflects the same pent-up frustration that helped inspire Dubin and Levine: "Our founders, Jeff and Andy, created Harry's because they were tired of overpaying for overdesigned razors, and of standing around waiting for the person in the drugstore to unlock the cases so they could actually buy them. When they asked around, they learned lots of guys were upset about the situation too, so they decided to do something about it."[12]

Raider implied in a 2014 interview that analyst reports about the men's shaving and personal care industry being "down and hurting" were misleading. "It's not hurting, it's moving," he said. "And it's moving in a direction that's great for us."[13] The market was also moving in a favorable direction for DSC.

In the summer of 2016, Dubin signed the documents for the sale of the success story he and Levine had cofounded a little over four years earlier.[14] Unilever, the consumer products giant and an archrival of P&G, had outbid several private equity firms to acquire DSC for an estimated $1 billion.[15] Such a transaction would mark a happy ending for any entrepreneur. But the disruptive work of the shaving clubs was not over. An analyst at Barclays commenting on Unilever's acquisition noted that "this is a bold statement that technology and product quality has reached a level where it doesn't matter as much as channel and business model."[16]

In the spirit of the Ends Game, had Gillette's decision to price the package of blades finally exhausted its capacity to generate revenue for the organization?

The Many Forms of Access Waste

The revenue model introduced by clubs such as DSC and Harry's to the sector tackles what we label access waste. Access waste occurs when the customer cannot readily obtain the means necessary to achieve a desirable outcome, or dispose of them when the outcome is no longer valued.

On the one hand, access waste can be traced to a *physical* problem in the purchase process, such a stockouts or inconvenience. Stockouts are the consequence of inadequate planning by customers or of storage limitations. We are all guilty of having run out of some product when we needed it most. Inconvenience results from too much time, too much distance, or too many process steps to overcome. Think of what happens when people do not have access to a washing machine in their home or building. In order to achieve the desired outcome of clean clothes, they are forced to invest time instead of money. They put a basket of dirty clothes in their car or under their arm, go to the laundromat, exchange bills for a handful of change or tokens, spend that cash on detergent, washing, and drying, and remain on the premises until the last item is dry. There is efficiency in the elimination of workarounds like this one, especially when people are not aware of their true costs.

Another, perhaps less obvious physical problem related to access is the unwanted accumulation of idle assets. This is the flipside of struggling to obtain a product or service. Sometimes customers want to "lose" access by getting rid of what they own but no longer use. This is not always easy, especially when the asset is relatively expensive such as a vehicle or fine clothing. The accumulation of idle assets has become so severe that people now spend $38 billion a year to stash what they own but don't need in self-storage facilities.[17]

On the other hand, access waste can also be traced to a *financial* problem. For instance, a business may lack the capital to purchase a product such as a fleet vehicle, a production machine, or a set of tools. Similarly, an individual may not be able to afford a car or a major appliance, either outright or even on an installment plan. Both business-to-business and business-to-consumer customers run the risk that they won't achieve their desired outcomes as they trade down on quality, find a workaround, or postpone or forgo the purchase. In the case of cars, manufacturers would struggle to increase access to their products solely by lowering the purchase price. While that may bring in more customers in the short term, such moves—whether promotional or permanent—can have dire financial consequences for an automaker.[18] Some customers who have the financial means to own a car may look for other options of transportation because they feel that purchasing one would be a misallocation of their resources. They may think that it is better to use public transportation, cabs, ride sharing, or other means rather than to own something that, on average, sits idle for the vast majority of its useful life.

A second financial problem related to access occurs when the revenue model prevents customers from achieving the variety of consumption they desire. The history of recorded music once again offers a good illustration, as it did in chapter 3. Prior to the digital age, the only way for customers to gain access to the music they desired, and the option to listen to it at any time, was to build a library of physical products. That became a prohibitively expensive undertaking for many people, especially toward the end of the twentieth century when customers needed to buy a full-length CD to get the one, two, or perhaps three songs they really wanted. One workaround to that access problem was illicit pirating, either on the private scale in the form of mix tapes and ripped CDs or on a grand commercial scale with free exchange of files on a service such as Napster.

The music industry currently solves that problem through popular streaming services, a model that tackles both physical and financial access inefficiencies by allowing listeners to have unlimited access to a library of music no matter where they are. The groundwork for the solution was laid step by step, first as a way to improve physical access. In 1997, Capitol Records made one of the first moves to bring music online when it offered a digital single ("Electric Barbarella" by Duran Duran) for sale as a download. Their internal digital team's proposal proved remarkably prescient and highlighted the access waste that digital music would reduce: "There is a group of people who like music but can't go to a record store anymore. It can be simply that their lifestyle doesn't permit it or that there is not enough time in the day. That group of people will purchase online and the music industry should find way to connect with them ... because the store is online, it's open 24/7 and it's global. With digital distribution, a product never goes out of stock."[19]

Over twenty years later, streaming services have not only taken hold, but also ignited a recovery in the music industry. In 2016, a report from Goldman Sachs referred to this new mode of generating revenue as "a massive game-changer ... that establishes a much more sustainable business model for the labels."[20] That forecast has held true, as the music industry's revenue growth and overall financial strength have improved significantly. By early 2019, Spotify reported that it had ninety-six million paid subscribers to its music service, an increase of 36 percent year-on-year.[21] The International Federation of the Phonographic Industry estimated that 255 million people around the world were using music streaming at the end of 2018, and they accounted for 47 percent of all revenue in recorded music.[22] The individual music labels are also benefitting. Warner Music Group reported record revenue for the 2018 fiscal year, stating: "We've had another terrific year and revenue exceeded $4 billion for the first time in our fifteen-year history as a standalone company."[23] The streaming services have become so pervasive that Apple decided in June 2019 to reconfigure its once-groundbreaking iTunes service. That initiative provided music fans access to individual, downloadable songs legally and at scale. But after eighteen years, its functionalities will be distributed across separate apps for Apple Music, podcasts, and TV.[24]

The Shaving Game Continues

What happened in the wake of Unilever's acquisition of DSC showed that shaving clubs would prove to be much more than a speed bump in Gillette's growth path. By introducing a new, leaner form of exchange with customers, the startups had permanently altered the decades-old trajectory of an entire industry. The new trajectory is still taking shape. In April 2017, Gillette made moves to "halt the inexorable surrender of its men's razor business to the newcomers."

First, the company cut prices. Under the headline "We heard you loud and clear," Gillette's websites explained the changes. "You told us our blades can be too expensive and we listened. You can find most Gillette blades and razors in the United States at lower prices without any compromise in blade quality."

But more important for the future of the business, Gillette introduced a new (to them) form of exchange with customers, the Gillette On Demand program, which offers some of the ease-of-use improvements and more affordable prices that DSC and Harry's pioneered.[25] Even the look and feel of the Gillette On Demand website shows many similarities to the approach taken by the disruptors.[26] Customers are "completely in charge. You can choose your own razor and how often you receive your replacement blades, making our shave club totally flexible to your individual needs."[27]

Gillette's use of the phrase "hassle of a subscription" is intriguing. It signals the possibility that the adoption of a revenue model that shifts the emphasis of the exchange between organization and customers from ownership to access is only a first step. Subscriptions are a hot topic as we write this book. But they are also only the first of many potential moves in the spirit of the Ends Game. The current shaving club models improve access by bypassing the "plexiglass fortress" in retail stores entirely. The convenience of the subscription model means that customers do not have to visit that section of the supermarket at all. It also means that consumption may increase, because the customer no longer faces the risk of running out of supplies. DSC explicitly recognizes and emphasizes how this convenience eliminates access waste when it states that it will ship its Restock Boxes whenever a customer wants so that "you never run out of anything you need to look, smell and feel your best."

But these changes may only mark a transitional step as the shaving companies learn more about the actions of their customers, moving them closer to understanding the value that shaving contributes to a desirable outcome we could phrase as "looking good and feeling good." What makes the difference, as we discussed in chapter 2, is information technology and customer impact data. The direct-to-customer model of shaving clubs helps each company close the loop on customer focus by providing rich, direct data on customers within a given community. To adopt the phrase of the analyst quoted at the outset of this chapter, access to impact data is today's "diamond-encrusted gold." The two-way flow of data between organizations and their customers has replaced "exposure" as the goal that every organization pursues. Such data are particularly important in the world of shaving, because both the frequency of consumption and the perceived quality of a shave can vary significantly from individual to individual.

When Unilever acquired DSC, the two companies stressed the start-up's "incredibly deep connections to its diverse and highly engaged consumers" and its "unique consumer and data insights."[28] The current communication strategy of DSC takes advantage of this opportunity to understand what customers want, and how these wants manifest themselves in individual buying behavior, which may change or fluctuate over time: "Tell us a bit about how you get ready, and we'll send you trial size products to dip your toes into the amazing waters of DSC. Two weeks later, we'll ship you a Restock Box with full sizes of all those products, at a discount, of course. You're in control. Any time you want, add and remove products, plus adjust how often you get Restock Boxes. See? Simple."[29]

Gillette also continues to collect, explore, and learn from customers. No two men or women shave in the exact same way, according to Kristina Vanoosthuyze, a principal scientist at the Gillette Innovation Centre. She elaborated on this variance: "There are guys that take 30 strokes and some take 700 strokes. Some people take 30 seconds, some take 30 minutes."[30] This knowledge can lead to more nuanced revenue models that come increasingly closer to reflecting the value that a company like Gillette has proven it can deliver.

The World of XaaS

The common denominator across all the access models described in this chapter is that revenue accrues to the organization as a function of time. For example, the basis of the shaving clubs' model is a periodic (monthly) purchase, even though they show product-based prices for cartridges, razors, or bundles on their websites. At first sight, this is nothing new. For decades, people have had subscriptions to newspapers, magazines, cable television, and so on. People pay rent. Retailers have long disguised their ownership-based approach to commerce by offering installment payment plans. But a host of recent technological changes—pertaining to monitoring, prediction, logistics, payment, and more—allows access models to spread across most sectors of the economy. Organizations today can lower the barrier of entry into a market by turning almost any good into a "service" with enhanced convenience.

Welcome to the world of XaaS—suggestive of the title of some forgettable science fiction novel from decades ago. In fact, the acronym XaaS, which is short for "Everything as a Service" implies that any transaction based on the transfer of ownership of a physical product (the X) can conceivably be replaced by a transaction offering access to that product on a periodic basis. XaaS has become the mainstream revenue model in software and tech industries, rendering the old perpetual license transaction comprising a software CD packaged inside a cardboard box obsolete. Citing an extensive survey of hardware and software buyers, the Boston Consulting Group notes that XaaS models "have become so pervasive and so desirable in the tech sector that cloud-like pricing is now a purchase criterion unto itself. Of the buyers surveyed, 77% said they would reallocate some spending or consider switching suppliers entirely if their current supplier failed to offer an XaaS model."[31]

The standard, time-based XaaS models effectively boost access in several ways. They create more discrete purchase and support options, including providing upgrades and training to customers. They reduce the financial burden on customers, especially for small and medium enterprises, by eliminating the need for the large upfront payment for a perpetual license. The cloud-based software giant salesforce.com expanded into the small- and medium-sized business sector exactly by attracting customers who were hesitant to purchase on-premise solutions.[32] Finally, XaaS models can lower search and administrative costs to practically zero. An executive at the messaging platform Slack captured that sentiment when he described his experience with Amazon Web Services (AWS), which provides "on-demand" cloud computing platforms to individuals, companies, and governments: "With traditional IT, it would take weeks or months to contend with hardware lead times to add more capacity. Using AWS, we can look at user metrics weekly or daily and react with new capacity in 30 seconds."[33]

XaaS models can also link together entire value chains that did not exist at the turn of the twenty-first century. Today, whenever someone brings up subscription models, it doesn't take long for that person to make a reference to Netflix. Yet in 2007, Netflix was a small but growing player in the DVD rental market, with a share of 12 percent and revenue of less than $1 billion. Then it launched what CEO Reed Hastings referred to as its "second act," an online movie rental service that did not require the customer to order, download, store, or remember to return the film or show.[34] Eventually, Netflix needed a type of subscription of its own in order to provide its streaming service to users: the company subscribes to AWS for access to cloud servers in order to "quickly deploy thousands of servers and terabytes of storage within minutes."[35]

Entering the Endless Closet, Garage, and Party

Cocktail dresses and evening gowns "sit idle" more often than even cars do. They serve a desirable and valuable purpose, but many are expensive and rarely worn, in some cases only once. Using the tagline "You don't have to own it ... to *own it*" Rent the Runway has a value proposition that is tailor-made to reduce the access waste in the market for designer clothes and accessories. The company offers "fashion freedom, a smarter closet, total wardrobe flexibility, and a smaller clothing footprint" at prices that are a fraction of the cost of one designer dress.[36] Jennifer Hyman, the company's cofounder and CEO, said, "Our goal is really to create the Amazon Prime of rental." A round of funding completed in March 2019 put the value of Rent the Runway at $1 billion.[37]

Rent the Runway allows women to select the clothes online, including seeing how the clothes look thanks to photos posted by previous users. Customers can choose a one-off rental, or subscribe to the basic service for $69 per month and receive up to four pieces at a time, including free insurance, dry cleaning, and shipping. But the advantages extend beyond having access to amazing clothes and accessories from at least 350 designers and lower prices. Women can receive personalized style tips and concierge service for fit and styling assistance.[38]

The disruptive effects of different revenue models in shaving, music, fashion, and other industries are no guarantee of success, however. In March 2017, General Motors launched the Book by Cadillac subscription service to test the viability of a Netflix model for luxury cars. For $1,500 a month, members could step in and out of Cadillac's ten models up to eighteen times a year without recurring additional charges.[39] Instead of buying one vehicle under the traditional model, subscribers essentially received temporary "ownership" of a vehicle and the license to use it and swap it in line with their personal needs, which can differ significantly for the same person or family. Taking a leisurely weekend drive, picking up friends or family at the airport, transporting bulky items, and rushing to a meeting in a crowded city all work best with vehicles of different shapes, sizes, and performance levels. Several other automotive brands, including Volvo, BMW, Mercedes-Benz, Audi, and Porsche, have launched similar subscription services.[40]

In December 2018, however, Cadillac temporarily closed down the Book by Cadillac service, citing some of the costs of operating the program, even at the small scale in only three cities. Those costs included the logistics of repairing damaged cars, cleaning them between uses and delivering them within twenty-four hours, as well as back-end technology issues that made some customer-service functions tedious and time-consuming.[41] The new version, scheduled for launch in 2020, will better integrate Cadillac's dealer network into the program.[42]

Another product that people run out of at inopportune times is liquor. When one finds an empty bottle in the cabinet, replacing it triggers a similar time-consuming, unpleasant process as in the case of razor blades. Solving this pain point is one of the many ideas behind what the French spirits company Pernod Ricard had in mind when it developed "Opn," a home entertainment system for serving cocktails. The system solves a host of access problems that people face when planning a party or simply mixing a drink at the end of a long day. It provides access to the spirits in the correct amounts, access to advice and cocktail recipes, and access to replenishments, all with the automated convenience. The design of the system also has appeal. Rather than a motley collection of bottles on a counter or a bar, Opn comprises a system of containers shaped similar to books—hence the name liquor library—on a docking station that integrates the book to mobile apps as well as a central database. The shape of the container is a key enabler for the collection of impact data. Each "book" stores a different type of alcohol, and the docking station tracks and reports how much of each book's contents remains, in much the same way your automotive telematics inform a driver about fluid levels (gasoline, water, wiper fluid) and air pressure. The docking station and accompanying software also analyze the data to determine whether and when to order replenishments, and also how many drinks one can still make from what remains.

"From helping put together shopping lists to seamlessly ordering spirits online and having them delivered to your door, creating social calendars and offering inspiration on the art of hosting, Opn will simplify and enhance the way we organize events at home, and help us prepare our favorite cocktails, in the smoothest possible way," Pernod Ricard described the system in a press release.[43] The four pillars of the system include the containers, the docking station, the app, and the website. One review picked up on the attractiveness of these features, noting that the tray or docking station "has an ability to track remaining levels of liquid and notify OPN system via the app (or a website) about the new orders as well as adjust cocktail recommendations based on what's available."[44]

In the introduction and chapter 3, we explained that the elimination of waste is cumulative. An organization cannot address consumption waste unless customers can access the product or service to begin with. However, some customers—whether they are a business or an individual—cannot guarantee that their rate of consumption justifies owning the product. Several cases in chapter 5 will describe solutions to this problem, as we look at consumption waste, its causes, and ways to eliminate it in the spirit of the Ends Game.

5 Flying Hours, Wash Cycles, and Miles Driven

Trevor Bauer, then a twenty-five-year-old member of the Cleveland Indians baseball team, gave his best effort to disguise an injury to his right hand during game three of the 2016 American League Championship Series. But after he had thrown only twenty-one pitches, the television cameras didn't need to zoom anymore to expose the evidence. Large drops of blood dripped from Bauer's pinky finger, and his manager removed him from the game.[1]

Bauer, who studied mechanical engineering in college and may be described as eccentric and nerdy, offered a rather exotic explanation for his injury.[2] A few days prior to that game, which his team eventually won, Bauer had gashed his finger and needed to go to the emergency room. The stitches he subsequently received could not withstand the pressure of gripping and throwing a baseball, and the gash reopened. When he initially injured himself, Bauer was not doing something mundane such as cutting vegetables in the kitchen or opening his fan mail with a pair of scissors. He was injured while repairing one of his unmanned aerial vehicles, which are more commonly known as drones.

Markets are slowly coming to grips with the vast potential, as well as the risk, that drones present. That risk makes insurance an important factor in how quickly the demand for drones and drone-powered services grows. The high-speed rotors from even a hobby drone can cause considerable physical injury, as Bauer learned the hard way. Critics and pessimists argue that drones can conceivably crash into power lines, damage buildings, collide with aircraft, and cause all kinds of other collateral damage. In late 2018 and early 2019, two incidents at important London airports underscore the havoc that drones can cause. First, reports of drone sightings forced Gatwick Airport to close temporarily, leading to the cancelation of flights affecting thousands of passengers. Two weeks later, authorities briefly shut down Heathrow Airport after another drone sighting. The *Financial Times* wondered whether airports can ever make themselves safe from drones.[3]

The challenge for an insurance company, then, is to understand and insure drone-related risk in an efficient way. That is, insurers need to offer coverage in a way that does not discourage organizations or individuals from using their machines as they would like. Flock, a startup based coincidentally also in the United Kingdom, feels it has the answer for commercial drone pilots as well as for hobby pilots such as Bauer. The company, which partners with German insurance giant Allianz as its underwriter, offers an on-demand, "pay-as-you-fly" product called Flock Cover.

What makes pay-as-you-fly different from conventional insurance policies is the way Flock provides the service and the way it managers risk. Technically speaking, the company offers "real-time micro-duration insurance" for drone flights. Pilots input information about their planned flight into an app (where and when they want to fly), which then displays data such as hyper-local weather conditions, ground hazards, and restricted airspaces.[4] An algorithm calculates and aggregates risk levels—using additional third-party information such as nearby aircraft, local topography (e.g., proximity to churches, hospitals, and schools), and roads and their current traffic levels—and determines a risk-dependent price for insuring that particular flight. The app also monitors the drone itself so that it can build a risk profile unique to that machine.[5]

At its core, the Flock story is about how a change in the revenue model can make commerce far more efficient, broadening the user base and increasing willingness to pay. Flock Cover enables pilots to undertake more flights, under more conditions, and make optimal planning decisions much faster. For the insurance provider, it represents less expensive access to more customers. More broadly, it provides them a way to enter a new market confidently and contribute to its rapid growth.

This "real-time micro-duration insurance" creates numerous efficiencies compared to a conventional policy, whose rates are based on the pooling of risk assessed mostly from historical data rather than the prevailing conditions for any given flight. As Flock CEO Ed Leon Klinger pointed out in a blog post: "These [pooled] risks are harder still for insurance companies to *quantify*—after all, insurers have no visibility into what pilots are actually doing with their drones, and the industry is relatively new (so there isn't much historical data to go on)." The result, Klinger wrote, was "outrageously priced (but often compulsory) insurance premiums for drone pilots, which can cost more than the drones themselves."[6] Flock offers policies that last for the duration of a flight, for a day, and has recently launched flexible monthly policies that allow "anywhere, anytime" coverage.[7]

Under the motto "the safer the flight, the cheaper the policy," Flock claims that pilots can save 15 percent on their premium per flight by using the app to optimize their flight planning. This 15 percent represents the average price difference between a pilot's first quote and the final risk-dependent quote for a given flight.[8] For hobby pilots, Flock offers coverage "from just £3 a day," or "the price of a coffee." The motivation is clearly to drive more drone usage: "The Flock Cover app ... takes under one minute to insure your drone flight; allowing you to take full advantage of a sudden sunny spell, or those perfect wind-free conditions."[9] Klinger feels that this same approach can apply to insurance policies for any future autonomous technological activity, from car trips to deliveries.[10]

In summary, the inefficiencies that Flock is trying to address are the prototypical causes of what we call consumption waste. Without a program such as Flock Cover, the purchase of expensive insurance cover can act as a deterrent to consumption or push the drone pilot (or company) to limit its use to certain environments. But an individual pilot or company ultimately wants to fly its drones. By providing access to reliable, affordable insurance at a moment's notice, Flock makes the use of the drone—its consumption—easier and more flexible. This encourages more and smarter consumption, because it provides customers with information on when and where to use a drone, knowing that they can assess situational risk. At the same time, the app offers its underwriting partner Allianz an inexpensive way to reach a larger audience. It is also advantageous to the insurance provider, because the algorithm determines the insurance coverage on the spot instead of waiting for a human being to process the information and make a judgment.[11]

At the same time, tackling consumption waste also puts the onus on Flock to have the insurance available whenever the customer needs it. This is the second, less obvious dimension of a revenue model that targets consumption. Because the customer pays only based on use, the firm clearly has the incentive to provide consumption "episodes" whenever they are demanded. The firm needs to be alert for occasions when customers demand more consumption or consumption under unusual circumstances, and make sure it has the capacity to meet this surge. When the item being consumed is insurance or some other intangible good, fluctuations in demand might seem to be less of a problem. A firm such as Flock may not "run out" of insurance, nor will insurance "break down" during operations as a machine could. But an insurer could conceivably run out of processing power to handle a sudden influx of requests. This underscores the link between access and consumption. An individual or business customer cannot consume something that is not accessible in the first place.

The Many Forms of Consumption Waste

Access is a question of scope: opening the market to as many (profitable) customers as possible. But even if customers enjoy complete, unrestricted access to a product or service, there is no guarantee that they take advantage of it. Indeed, consumption waste can take on many frustrating forms. It occurs when an asset—a car, an apartment, a bicycle, a printer, a medical device—either sits idle for a large portion of its useful life or is not acquired in the first place because customers believe that the expected use does not justify committing to owner-ship. In chapter 3, we mentioned that automobiles sit idle about 95 percent of time, which is a disturbingly low level of utilization for such an expensive product. However, the same fate applies to a large range of everyday durable goods and specialized business equipment.

As is the case with drone insurance, consumption waste also occurs when some barrier—such as a related risk or the price of a comple-mentary good—prevents someone from using a product or service they already have access to. Finally, consumption waste results when cus-tomers are forced to buy a package that is either too small or too large given one's need, and therefore end up forgoing additional consump-tion or throwing away any unused portion.

The three options available to organizations to tackle these con-sumption problems are *unbundling, metering*, and *sharing*. Each of these strategies, in its own way, improves efficiency by removing barriers and by activating dormant or underutilized capacity. Each strategy can trace its origins back decades or even centuries, but the twenty-first century's ongoing advancements in information and communi-cation technology—and the transparency those advancements have fostered—have expanded their potential. Each option also differs in terms of complexity, required investment, and upside. Unbundling is the most straightforward approach, followed by metering. Sharing, which today is perhaps most popular in markets for transportation and lodging, is the most complicated.

How Unbundling Eliminates Consumption Waste

In chapter 5, we referred to Netflix and Spotify as examples of libraries: large collections of similar products (movies, television shows, songs) that would be prohibitively expensive for customers to compile on their own. Rent the Runway plays a similar role with designer clothes and accessories. Cable companies, however, are more like uber-libraries because their service bring together individual libraries (the different channels) as well as Internet access and fixed-line telephony as part of a "triple-play" package. Cable service has become so comprehensive and bloated that users end up paying for libraries they want and for libraries they don't want. For example, some people may spend the bulk of their time watching live sports and 24/7 news but rarely watch children's programming and documentaries. Other subscribers may do the exact opposite.

This is an example of the consumption waste that occurs when customers are forced to buy a "quantity" that does not correspond to their actual needs. While subscribers technically do not throw away the excess movies, shows, events, and documentaries they pay for in the subscription, they are likely to consider the money spent as thrown away. In response, viewers have undertaken an action known as "cord cutting." They are dropping their subscriptions under the assumption that it is more efficient for them to build their own bundle, including only the content that they are really interested in. The prices for the individual "libraries" support this rationale. A recent study conducted by Consumer Reports, a leading consumer watchdog organization in the United States, revealed that, across all providers, the median price for a cable service bundle was $173 per month.[12] This is a far cry from the $15.99 per month that, as of June 2019, Netflix charges in the United States for a premium plan.[13] Similarly, YouTube TV offers "cable-free live TV" for $49.99 per month and HBO charges $14.99 per month.[14] The incredible growth of these services suggests that people are proactively trying to align their spending with the kinds of programming choices they prefer, thus reducing or eliminating consumption waste. A 2019 report by PwC forecasts that the revenue of streaming video services will increase by 64 percent over the next five years to almost $24 billion per year, while cable and satellite TV providers will see revenue decline by 16 percent to $84 billion in the same period.[15]

Meanwhile, the evolution of the recorded music and newspaper industries over the last twenty years demonstrates just how intertwined access and consumption waste can be. It exemplifies what happens when a sector transitions from selling physical goods to selling digital ones. Physical constraints made it economical to bunch songs onto a compact disc and articles into a newspaper. The most efficient way to distribute the most news to the most people used to be the printing press. But in digital form the economics change. It is just as efficient to supply individual songs and articles, introducing the possibility to "package" the good in whatever size best fits the consumption pattern of customers. As Netscape cofounder and Internet pioneer Marc Andreesen said, "Bundles emerge as a consequence of the current technology."[16] Cycles of bundling and unbundling are a natural consequence of the advances in technology that we described in chapter 2.

In the specific case of recorded music, the rise of Napster and peer-to-peer sharing at the dawn of the twenty-first century established that listeners preferred to purchase individual songs rather than complete albums. Yet at that time the record labels earned the vast majority of their revenue from selling shrink-wrapped jewel cases containing plastic discs with a fixed set of songs on them. It didn't matter whether listeners wanted all the songs or just one or two. Ultimately, the demise of the compact disc led to industry-supported systems for the sale of digital tracks, triggering the explosive growth of iTunes. By 2013, iTunes had sold twenty-five billion individual digital tracks, all unburdened by the unwanted tracks that would have come with purchasing the traditional album.[17]

How Metering Eliminates Consumption Waste

What is the household chore people hate the most? According to a 2018 report by the Council of Contemporary Families, it's doing the dishes, a task that one of the report's authors described as "gross. There is old, moldy food sitting in the sink. If you have kids, there is curdled milk in sippy cups that smells disgusting."[18]

If the task is bad enough at home, imagine doing dishes on a much larger scale, such as in a hotel or restaurant. Then imagine the demanding quality requirements for washing the dishes and silverware at elegant venues such as the Opera House in Sydney or luxury hotels such as the Mandarin Oriental in Beijing or the Burj al Arab in Dubai.[19]

That kind of work is the specialty of the German company Winterhalter, which specializes in "warewashing" equipment and services for commercial kitchens. Founded in 1947, the family-owned company offers its customers the dishwashers, racks, detergents, water treatment supplies, and services they need to clean their dining ware. But what makes Winterhalter truly stand out in the market is how it charges its customers for those clean dishes. Instead of relying on the simple, traditional model of selling the dishwashers and consumables, or an access model such as the subscriptions we described in chapter 4, the company went a step further and launched a program called Pay per Wash. In the spirit of the Ends Game, Pay per Wash focuses on eliminating the inefficiencies that hinder consumption, which by definition also includes problems of access. Winterhalter stresses that this program leaves the customer with "no investment, no risk, and no fixed costs."[20] The customer logs into the Pay per Wash portal, enters the number of wash cycles according to its preconfigured specifications, and starts the machines. Winterhalter then charges for each completed wash cycle. It claims that it "boosts commercial warewashing to the next level: the first warewashers you only pay for when you actually use them. That means perfect wash results for everyone—regardless of the available budget, and with maximum flexibility."[21]

In chapter 4, we showed how some companies have begun to adopt XaaS subscriptions or memberships in order to minimize access waste. By substituting outright ownership with periodic payments, these revenue models aim to remove the physical or financial obstacles that prevent some customers from "reaching" a desired product or service. However, while a well-designed access model can truly open up the market, it does not track the consumption that is necessary for customers to derive value. This is the second checkpoint we referred to earlier. Rather, the products or services to which customers may subscribe on a monthly or yearly basis are rigid, predetermined bundles.

Winterhalter eliminates this rigidity by metering and charging for discrete uses. This approach is also a superior way to account for the symbiotic relationship between industrial products and the services that accompany them. Think of the ways that manufacturing companies typically charge for services. When a manufacturer bundles its "free" services into the price of its product, this effectively means that it charges for them by the pound, square foot, or some other metric that has nothing to do with their underlying nature or quality.[22] This simple example underscores how bundling products and services together can distort customers' views of what they are paying for.

Charging for products and services separately sidesteps the service-by-the-pound issue, but creates other problems. If customers do not calibrate their service needs properly, or cut corners due to cost constraints, they face the risk of owning assets that fall short of their goals for throughput, efficiency, or running time. Sticking with these traditional models can result not only in a considerable amount of access waste, but also in consumption waste. It can turn even the most customer-friendly, high-quality suite of services into an impediment to a company's growth and sustained success.

The success of Winterhalter's Pay per Wash, and of any revenue model anchored on consumption for that matter, is clearly tied to the reliability of the product. This point is perhaps obvious, but no less important. If the dishwasher does not work properly when the customer presses Run, Winterhalter does not get paid. Analogous to what we said with respect to Flock and drone insurance, the onus is on Winterhalter to make sure its products are in perfect working order when customers need them. The customer pays only based on actual consumption, which gives the firm the incentive to ensure that consumption "episodes" happen when demanded.

Yet Pay per Wash might only be a prelude to an even leaner revenue model from Winterhalter, which is currently developing what it calls Next Level Solutions. The goal is to use Internet connectivity to optimize the efficiency of its dishwashers. By collecting and analyzing impact data from individual customers, Winterhalter hopes to come ever so close to understanding and quantifying cleanliness, and then its value to the user. Ralph Winterhalter, grandson of the company founder and currently joint CEO, described the initiative, stating that "exploiting technological advances is the way to secure our brand for the future and ensure we enjoy sustained growth and expansion."[23] It brings the company yet another step closer to fulfilling its goal of "assuming responsibility for a perfect cleaning result."[24]

Switching sector, think now of the mobile data plan a household might sign up for. The mobile network operator counts the number of gigabytes, not the number of times a family member uses a particular app. On a much larger commercial scale, think instead of a software package available to all the employees of a client organization. Some employees may be heavy users, others may be moderate users, and some may rarely use the software at all. The developer in this case might track the time spent logged into the system, data consumption, and other parameters across all users.

These two situations also lend themselves to a revenue model where the organization charges customers based on the actual consumption of the product or service it brings to market. In 2018, Deloitte conducted a survey with more than one thousand information technology specialists at large U.S. companies in order to glean their attitudes toward metered XaaS subscriptions, which focus on use, compared to the traditional purchase of a perpetual license. The specialists cited many advantages of these newer agreements, which are often referred to as flexible or on-demand consumption, with operational and workforce excellence at the top of their lists. A large majority of respondents also mentioned the reduction of time spent on maintenance and upgrades, responsibility for which now shifts to the supplier, and the opportunity to access the latest technologies.[25] The survey found that suppliers that have introduced metered XaasS subscriptions "have been rewarded by consumers as well as investors, challenging conventional valuations and placing pressure on traditional industry players that are retaining traditional business models."[26]

Interestingly, these findings apply not only to the latest innovations coming out of Silicon Valley, but also to classic product categories that are more than a century old. One customer who agreed to a metered XaaS contract in one of these traditional categories said

the "major benefit to us is financial planning" and that the arrangement "has taken away the headache" of routine maintenance checks.[27]
Another customer of the same supplier is convinced that the metered
XaaS model is "the correct strategy for our business" because of the
economies of scale.[28]

Again, these customers were not buying software, computer hardware, or some other modern technological wonder. They were buying
Michelin tires for their commercial vehicle fleets. To be more specific,
they were signing up for the EFFITIRES program offered by Michelin
Solutions, the fleet management unit of the giant French tire manufacturer. Michelin supplies the tires, but the client fleet managers pay only
for miles driven. The company stresses that the program will "protect
your drivers and their loads, while still minimizing your total cost of
ownership."[29]

One analysis published by the World Economic Forum claims that
Michelin's shift from "selling tires as a product to a service" has helped
the company "achieve higher customer satisfaction, increased loyalty
and raised EBITDA margins."[30] Michelin has built on that success by
expanding the program to other types of commercial vehicles, using
different metrics based on use. The company says it has "full solutions
available for professionals in any industry" with models based on "kilo-
meters driven, number of landings for airlines, [and] tonnes transported
by mining sector customers."[31] The elements of these "pay-as-you-go"
programs include tire selection, mounting, maintenance, assistance,
regrooving, and end-of-life recycling.

Since 2015, the program also includes a fuel efficiency commitment
for commercial vehicle fleets. In 2017, Michelin acquired NexTraq, a
commercial fleet telematics company. At the time, Michelin stated that
"telematics and fleet management services are a rapidly growing category worldwide and an important area of Michelin Group's overall
business plans."[32] In 2018, the company highlighted its efforts to boost
the collection of impact data at the individual product level as a means
to improve its tire management services. Ralph Dimenna, a Michelin
senior vice president said, "It's a matter of collecting enough data on

that tire that the customer understands the [value] of that asset." He added that the data will also enable Michelin to shift from reactive to preventive and even predictive tire management.[33]

Finally, one rapidly expanding group of demanding customers in the twenty-first century are the digital nomads, people who are "extended travelers who work remotely with the help of digital tools like a laptop or smartphone."[34] Whether they are traveling the globe or within their own region or city, these workers relish their freedom and their ability to work anywhere. But working anywhere ultimately means working somewhere, and while beaches and mountaintops have an undeniable appeal, many of these nomads gravitate to practical places such as restaurants or coffee shops that offer the right mix of comfort, convenience, caffeine, ... and WiFi!

The Russian café chain Ziferblat uses a metered model to cater not only to digital nomads, but also to anyone who wants to find a place to socialize, play games, or simply chill out. Unlike other coffee shops or gathering places, the food, beverages, restrooms, and WiFi are free of charge. Instead, patrons pay only for the time they spend at the premises, measured in minutes. Ziferblat—which means "clock face" in Russian—opened as a "pay-what-you-want" room for poetry readings back in 2010, but quickly switched to the pay-as-you-go model.[35] By 2015, the company had thirteen locations in Russia, Slovenia, the United Kingdom, and Ukraine. The café in Manchester, England opened in 2016 and attracts ten thousand guests per month.[36] As one Ziferblat executive explained, "We don't judge, we just ask people to respect the space." Some guests come only to use the bathroom or grab a quick bite to eat, while others spend all day getting work done.

The Ziferblat model offers some food for thought (no pun intended!) for other gathering places where the value patrons derive has some correlation with the time spent on the premises. These include the popular board-game cafés such as Snakes & Lattes, which operates three locations in Toronto as well as one in Tempe, Arizona.[37] For a small cover charge per person, guests can stay as long as they like and play board games selected from the location's extensive library, which includes more than a thousand titles. If the group of customers is struggling with the rules, they can ask an onsite "game guru" for an explanation on how to play.[38] The appeal of the current model is that it eliminates access waste, because assembling a large library of board games is prohibitively expensive for most people, and opportunities to try out a new game comfortably with friends and with expert help are rare. The next step would be to identify consumption waste by creating incentives for guests to come more frequently, play more often, or consume more on the premises.

The successful evolution of revenue models from ownership to metered service in anything from hardware and software to established industrial products and hospitality begs an important question: When does it make sense to charge for something you are tracking? In the spirit of the Ends Game, the answer to that question will always depend on whether the switch makes the exchange between the selling organization and the buying customer more efficient. To what extent does the change in revenue model eliminate consumption waste by making it easier, more cost effective, or more valuable for the consumer to use a product or service? To what extent does it improve the utilization of the underlying asset?

How Sharing Platforms Eliminate Consumption Waste

In chapter 4, we explained how Rent the Runway addresses access waste in the market for fashion by offering women the opportunity to enjoy a wide variety of dresses, gowns, handbags, jewelry, and other designer clothes and accessories without having to purchase any of these items outright.

Violet Gross and Merri Smith appreciate Rent the Runway, but the two entrepreneurs thought they could take the model one step further. Instead of a service that rents clothes from a curated "closet" stocked with popular brands, they envisioned a service that builds on something women have always traditionally done: borrow clothes from other women.

"My friends and I are always borrowing each other's clothes," Smith said in an interview in October 2018. "Violet and I realized there's an untapped opportunity here. A business of renting friends' clothes can have serious legs with the right infrastructure."[39] That infrastructure became the app Tulerie, which Gross and Smith launched in 2018. In the spirit of the Ends Game, the Tulerie website is clear that the business addresses the consumption waste inherent in a traditional ownership model: "Tulerie is an opportunity to both expand your wardrobe options and profit on what you own by sharing closets with conscious and fashionable women in your city and around the U.S.... By borrowing versus buying, you can indulge in of-the-moment pieces you would never wear to its fullest potential alone. By lending, you can underwrite those heritage pieces from high-end designers (because ethical is expensive) by temporarily sharing it with women just like you."[40]

Uber and Airbnb serve as shorthand examples for the sharing economy, also referred to as collaborative consumption, but the Tulerie example shows how one of the archenemies of efficiency—the idle asset—lurks literally in any market where a significant number of customers, individuals or businesses alike, own things they aren't fully using, and a correspondingly significant number need exactly the same product or service but cannot afford to purchase it. The idle asset is perhaps the clearest symptom of the consumption waste prompted by the classic "pay-to-own" model of generating revenue.

Sharing platforms provide access to products and services to more people. They also increase the use, and thereby the return on investment, of assets that individuals already own. This logic applies as much to real estate, where Airbnb matches travelers with open homes and apartments, as it does to warehouse space, where logistics startup Flexe matches retailers with depots that have excess capacity. It also applies in the "market" for parking spaces, where platforms such as Spothero help drivers find open spots in crowded cities by pooling the excess capacity of its participating partners.[41] Finally, the logic of collaborative consumption applies to services such as TaskRabbit, which helps people find available labor to do odd jobs such as mounting a flat-screen television, assemble furniture, or move heavy objects. In this case, there are idle assets on both sides of the exchange because, for example, customers cannot make use of unassembled furniture and those offering the labor have some spare time on their hands.[42]

While the number of sharing platforms is increasing rapidly, perhaps the biggest impact from their presence is felt in less-developed, rural economies where communities have traditionally shared assets out of necessity, but only at a familial or local level. Modern information technology empowers the sharing of critical assets such as farming equipment on a much larger scale, increasing the well-being of both equipment owner and equipment users. Trringo, India's foremost tractor and farm equipment rental service, is a great example. Trringo "aims to raise the level of mechanization in farming through the power of technology and a strong franchise network to make farm mechanization easily accessible, affordable and reachable to farmers across India."[43] This statement underscores the nature of the relationship between initiatives to reduce consumption waste and initiatives to reduce access waste. While access to farm equipment does not guarantee consumption, there is no consumption without access.

How large will the sharing economy become? PwC has estimated that the transaction volume across just five types of platforms in Europe—collaborative finance, peer-to-peer accommodation, peer-to-peer transportation, on-demand household services, and on-demand professional services—could reach €570 billion by 2025, up from just €28 billion in 2016.[44] If the forecast becomes reality, the estimated revenue for the platform providers in these five markets could reach €83 billion, up from €4 billion in 2016. In China, the growth is similarly staggering. China's Sharing Economy Research Center (SERC) estimated total transaction volume in 2018 at roughly $440 billion, with an annual increase of 42 percent.[45] The SERC's deputy director, Yu Fengxia, expressed optimism for future growth in terms that fit hand in glove with the spirit of the Ends Game: "The sharing economy's potential of stimulating consumption will be released, as it can not only satisfy consumers' needs limited by the traditional service mode, but also boost their new consumption needs."

Li Xiao, founding partner of the Chinese venture capital firm Joy Capital, added, "We are living in a society which constantly pursues efficiency improvement. Resources that are not fully utilized give birth to sharing economy. I am confident that as the sharing economy penetrates into more and more industries, it will create great value to our society."[46]

The growth of these revenue model in any region of the world will depend on whether and how the platforms and their participants— asset users and asset owners alike—overcome established social norms that frown on sharing. A survey among European consumers conducted by the Dutch financial group ING showed significant lingering aversion to use collaborative consumption.[47] In the Netherlands, 64 percent of respondents said, "I don't like other people using my property." Concerns about insurance was a deterrent for 44 percent of the respondents, and concerns about the quality of the shared items discouraged 32 percent of respondents. Yet history shows that social norms do change and taboos can be overcome. Think about the various ride-sharing services. Most members of Generation X or older generations can recall their parents admonishing them to "never get in a car with strangers" or warning them about the dangers of hitchhiking. But outfits such as Uber and Lyft are essentially large-scale, organized, peer-reviewed hitchhiking services. The difference is that today people use their thumbs to tap their smartphone screen instead of holding them out to signal they need a ride.

Organizations that anchor their exchanges with customers on consumption improve on the alternatives of selling ownership or targeting access alone. At the same time, a "pay-per-use" arrangement may still be inefficient if consumption fails to provide the performance customers seek from a product or service. Accordingly, the next and final checkpoint in the journey to lean commerce is a revenue model that focuses on outcomes or, in some cases, on the ultimate outcome: value itself. This is the subject of the chapter 6.

6 Laughter, Rocks, and Quality of Life

The Spanish government once risked throwing the country's local live entertainment industry into a tailspin by raising the tax on tickets to theater performances. When the tax rate skyrocketed from 8 percent to 21 percent in 2013, the country's small theaters—never blessed with stellar finances anyway—scrambled for ways to stay afloat.

To keep revenue flowing, Teatreneu, a popular comedy theater in Barcelona, Spain, decided to offer spectators a novel proposition. The logic behind this proposition was simple: if Teatreneu is in the comedy business, then it should sell comedy, not tickets. Accordingly, Teatreneu introduced a groundbreaking scheme that became so effective and so popular that it not only boosted attendance and average revenue per spectator, but also saw its app and advertising campaign garner international attention and earn major global marketing awards.[1]

The system was called Pay per Laugh.[2] Spectators entered the theater free of charge. A facial recognition system mounted on the back of the seat in front of them registered each time they laughed during the performance. Each laugh was priced at 30 Euro cents. Teatreneu set the maximum charge at 24 Euros per show, or 80 laughs, so that "no one would need to cry because they laughed more than they could afford."[3] This decision effectively created a spectrum of enjoyment from "no fun at all" to "non-stop fun." The unlucky spectator who hated the performance—or at least never laughed—paid nothing. Conversely, the lucky spectator who had non-stop fun paid the maximum amount, which was still reasonable. In between those two extremes, the number of laughs provided a clear, consistent, and quantifiable—albeit still imperfect—way to measure entertainment.

By bringing its revenue model into better alignment with the satisfaction of spectators, Teatreneu virtually eliminated what we label performance waste. Performance waste occurs when a product or service in the market is accessed by customers, it is used or experienced, but it doesn't deliver the value expected from it. Perhaps the most prominent feature of revenue models that tackle performance waste is that they shift all the responsibility for a successful exchange from the shoulders of customers to those of the organization. This is certainly the case with Pay per Laugh. The quality of the show matters because it determines the financial return to Teatreneu. Yet an extremely poor performance doesn't cost anything to the audience, except perhaps the time wasted in the theater.

At the same time, an organization that uses a performance model lives and dies by the "quality" of the metric it adopts. Some people may enjoy the show immensely but laugh very little, while others may attempt to stifle laughter in order to save some money. These concerns are always going to exist unless the metric is a perfect, tamper-proof proxy of the actual value derived by customers. Finally, the right technology is essential to make pay-by-outcome work. The practice of selling entry tickets to shows is centuries old and could hardly be less sophisticated. The Globe Theatre charged admission to performances of Shakespeare's plays in the 1600s, and even the Colosseum in Ancient Rome had a seating chart.[4] Teatreneu's Pay per Laugh would be little more than wishful thinking without facial recognition applications, the hardware to deploy them across the theater, and the software programs that tally the laughs and allow spectators to share their experience on social media. Subjective outcomes such as one's enjoyment are difficult to measure, but companies are increasingly finding novel solutions thanks to technologies that were underdeveloped or even nonexistent as recently as ten years ago.

Value Is Elusive

Value is the ultimate outcome. If a firm could charge its customers based directly and precisely on the tangible and intangible satisfactions they derive in an exchange, then there would be no need for an intermediate measure to calibrate the exchange and access, consumption, and performance waste are minimized. Value establishes the natural equilibrium between *You get what you pay for* and *You pay for what you get*.

But finding a way to capture value consistently and reliably is often an insurmountable challenge. The practical alternative is to settle on a proxy: an outcome that can be quantified and verified and, importantly, is an accurate representation of value. "Enjoyment" is a good example of value that, at first sight, defies consistent quantification. The standard pay-to-own revenue model of theater companies is clearly inefficient because ticket prices are set before a show and fixed for all spectators. To reduce access and consumption waste, one could replace the ticket with a pay-per-time model, similar to the earlier example of Ziferblat. Yet comedy is ultimately a matter of taste, and metering here implies that spectators who stay in their seats for the duration of the show pay the same, regardless of whether they actually enjoyed the performance or not. To pay less, disappointed spectators would need to stand up and leave early. Which brings us to this question: How can a business register someone's "good time" with confidence? Better yet: How can that business come up with a system that charges customers based on the amount of good time they had? Teatreneu's solution was to use laughter as the proxy for enjoyment.

Most industries are at the early stages of a process that will unfold over the next few years, with improving technology making performance models not only feasible, but also practical and profitable. The primary concern for organizations in the meantime is understanding the true source of the value they create for customers—often with the collaboration of partners and customers themselves. If value itself cannot be measured, the choice of outcome is critical. There may be factors that contribute to an outcome that the organization cannot observe, measure, or control. To the extent that there are significant differences in the value customers derive from a product or service, then the chosen outcome measure must be "personal" enough to reflect this.

In the remainder of this chapter, we will look at how other sectors or individual firms define and employ outcome measures in their performance models. In some cases, the proxy is already established in the industry. In other cases, the debate is ongoing.

When Value Comes from Mood

How much do people enjoy the music they listen to? Does their appreciation for particular songs peak and then diminish over time, or does it persist regardless of the number of times a given track is played? Importantly, can one's choice of music influence downstream behaviors and create value by enabling or preventing certain outcomes?

One group of academics has examined how existing technologies can generate music recommendations for drivers based on their mood. They argued that "considering the high correlation between music, mood, and driving comfort and safety, it makes sense to use appropriate and intelligent music recommendations based on the mood of drivers in the context of car driving."[5] In the spirit of the Ends Game, such a statement begs the question of what is the right outcome measure. One official at Ford believed that the proper proxy is the minimization of "driver distraction and stress."[6] As far back as 2013, Ford was working with software and hardware developers to explore and capitalize on "unheard-of access to vehicle data, entirely new application categories and hardware modules" that can promote "safety, energy efficiency, sharing, and health."[7] The other interesting questions are how to measure the driver's mood, which is a critical ingredient, and what type of performance model can turn these impact data into revenue. One idea could be that a music streaming service "re-imagines itself and its relationship with consumers," as the market becomes both saturated and more competitive.[8] For example, drivers could pay based on the pleasure they derive from listening to music. These data could in turn influence insurance premiums, prices at highway tolls, and other driving-related costs. Drivers could be rewarded if there is indeed a correlation between listening to music and safer driving.

Specifically, automakers and their tech partners are exploring three sources of information to understand mood behind the wheel: external factors such as weather, traffic, and road conditions; telematics data from the car itself; and biometric data collected directly from drivers. In the future, cars could use a combination of biometrics such as facial recognition and vital signs such as heart rate, breathing rate, and even sweat to measure the stress level of drivers. One company is even experimenting with a gel applied to the steering wheel that can serve as a biometric sensor.[9] When drivers get into stop-and-go traffic, an intelligent radio can shift to a more mellow station or playlist because it "knows your braking and acceleration patterns."[10] The idea behind these initiatives is that the change in music might not only provide entertainment, but also lower the risk of road rage and, hence, lower the risk of accidents. More than 90 percent of automobile accidents involve human error, therefore knowing and "adjusting" the mood of drivers may be a way of limiting problems with severe consequences.[11]

When Value Comes from Clicks

The field of advertising has long faced the challenge of defining meaningful outcomes. Prior to the Internet, it was difficult, if not impossible, to say with any certainty whether someone in the target audience saw or heard an advertisement, never mind engaged with it or acted on it.

Then along came the Internet and, with it, a pioneering company that wanted to inject accountability in the process of monetizing advertisements. "The whole point of Internet advertising, I thought, was accountability," said one of the founders. "You could measure it, unlike with print ads. But here was everyone still selling ads the old way: buy a bunch of impressions, cross your fingers, and hope it turns out well."[12] Accordingly, this newcomer ranked search results not by where and how often keywords appeared, but by how much advertisers were willing to pay for them. The company "posts the per-word pricing in an open auction, allowing Web sites to continually bid for higher placement on a given topic."[13]

You would be correct in thinking that this innovation resembles what Google does today with its Google Ads platform. In fact, this concept first appeared in a *New York Times* article in March 1998, about six months before Google even came into being. The pioneering company was GoTo.com (which in 2001 renamed itself Overture Services and in 2003 was acquired by Yahoo!), and the founder quoted above is Bill Gross, not Larry Page or Sergey Brin. Google first launched its own auction-based, pay-per-click search-advertising product in 2002.[14]

Today, Google claims that its customers "only pay for results, like clicks to your website or calls to your business."[15] More specifically, Google offers "advertising on a cost-per-click basis, which means that an advertiser pays us only when a user clicks on an ad on Google properties or Google Network Members' properties or when a user views certain YouTube engagement ads."[16] This is another quintessential example of lean commerce, and one that fulfills the desire for accountability that Gross expressed. Advertisers pay only for a specific, desired outcome of advertising rather than purchasing advertisements and hoping that the audience engages with the content. Google's model has become so successful that its parent company Alphabet's revenue from advertising totaled $116.3 billion in 2018.[17]

Looking ahead, one may question whether clicks is the best proxy of actual value. Performance waste is a matter of degree, and while the ability of an advertisement to drive online traffic is important, traffic is not the same thing as purchases. To the extent that advertisers ultimately care about actual transactions on their websites, a better performance model today may be "pay-per-conversion," where advertisers pays Google or one of its competitors only when users click on an advertisement and then shop. Whether pay-per-conversion ultimately replaces pay-per-click depends in great part on the development of measurement technology that can establish the causal relationship between exposure to an advertisement and purchase behavior accurately and transparently.

When Value Is Life Itself

Despite years of almost constant experimentation, the use of outcome-based agreements in the health-care sector is still in its infancy. The examples that follow demonstrate just how difficult it is within the sector to represent "health" in a manner that is consistent, reliable, and, above all, accurate. The challenge in the United States is particularly acute, with spending seemingly out of control. For instance, a 2017 article in the Washington, DC, policy and politics journal *The Hill* claims that, while around 90 percent of all drugs on the market are low-cost generics, "roughly 5 percent of patients take so-called 'specialty' drugs to treat serious or life-threatening diseases. These drugs represent one-third of all drug spending, and this trend is expected to continue with the discovery of new treatments for rare diseases and other highly personalized medicines."[18]

One existing and relatively popular measure is the quality-adjusted life year (QALY), which is calculated by estimating the years of life remaining for a patient after a particular treatment or intervention, weighted by a quality-of-life score on a scale from 0 to 1.[19] Several countries, including Canada, the United Kingdom, Ireland, and the Netherlands use QALYs as the starting point to calculate the value of a specific drug or treatment and to judge what the burden on public funds should be.[20] But the adoption of QALYs is more controversial in the United States and explicitly forbidden under some circumstances by the Affordable Care Act (ACA).[21] One justification for the ban is the clear moral dilemma that QALYs expose: "This is America not wanting to put a value on the price of a life."[22]

Roche, a Swiss multinational, is developing Personal Reimbursement Models (PRMs), which consider that the effects of medications typically vary by indication (a patient's specific condition), combination (with other medications), and response. This approach clearly departs from the tradition of charging for a pill or treatment—the legacy pay-to-own model in the industry. Under a PRM, the price is "driven by the value the therapy delivers to patients, and one product can have different prices."[23] Drug utilization data help provide the necessary insights into efficacy. Roche believes that PRMs "will accelerate patient access to innovation and reduce financial pressure on prescribing by enabling the benefit of a medicine to be better reflected in its price."[24]

In 2007, Johnson & Johnson proposed a pay-by-outcome model for an oncology treatment in the United Kingdom, under which the company would refund any money spent on patients whose tumors did not remiss. That same year, Cigna, a major insurer in the United States, suggested that makers of cholesterol treatments (statins) pay the medical expenses of patients who suffered heart attacks, even if they had complied steadfastly with the treatment regimen.[25] Finally, in 2017 the drug maker Amgen and health insurer Harvard Pilgrim reached an agreement similar to the one Cigna suggested. The contract provided Harvard Pilgrim with a rebate if an eligible patient has a heart attack or stroke while on Repatha, a treatment intended to reduce the risk of heart attack or stroke by lowering LDL (bad) cholesterol.[26]

While these arrangements are often referred to as "pay-for-performance" or "value-based" schemes, fundamentally they couple a traditional ownership model with a money-back guarantee. As we mentioned at the outset of the chapter, a prominent feature of pay-by-outcome models is that they shift the responsibility for a successful exchange from the shoulders of customers to those of the organization. This rebalancing of risk is one of the primary mechanisms for reducing or eliminating performance waste. But the timing of payment matters. Even if patients are promised a refund in situations where the desired outcome does not materialize, they still face the upfront expense and some uncertainty as to whether they will actually get their money back, which together reduce the efficiency of the exchange.

In 2018, the year after it reached the deal with Harvard Pilgrim, Amgen reduced the list price of Repatha by 60 percent to $5,850 exactly in response to these contingencies.[27] Amgen chairman and CEO Robert A. Bradway explained the move by saying: "Concerns over out-of-pocket costs have proven to be a barrier to its use for too many patients. We want to make sure that every patient who needs Repatha gets Repatha."[28]

When Value Is a Pile of Rocks

Some twenty-first century businesses have an air of science fiction to them. They are an intricate combination of information technology and intimate customer focus. Using sensors, digital platforms, cloud computing, and machine learning they understand the varying circumstances of customers, then draw on their experience and technological expertise to create superior solutions to their specific needs and wants.

One might think that this description is more likely to fit a clean, "sophisticated" industry such as robotics or autonomous vehicles than the grubby type you may encounter on the television series *Dirty Jobs*. Yet the Australian company Orica, the world's largest provider of commercial explosives and blasting systems to mining companies, exemplifies how a smart choice of revenue model is critical to sustained success.

Instead of selling explosives and blast-related services, Orica makes a living based on the quality of the "broken rock" it delivers. This outcome-based model, which they refer to as Rock-on-Ground contracts, has become a defining characteristic of how Orica manages the business, both internally in terms of innovation and product development, and externally in terms of its ongoing relationships with customers and positioning in the market. The size of the broken rock that results from a blast appears to correlate well with the value mining companies derives from using explosives, as smaller rocks are easier and cheaper to dispose of.

Think back to chapter 3, where we cited Theodore Levitt's insight that customers "don't want to buy a quarter-inch drill, they want a quarter-inch hole."[29] Continuing this example, instead of selling customers the equipment to blast a "hole" on the mine surface, Orica ultimately focuses on the complete act of creating that hole, from the planning stage through to the aftermath. Citing academic research, Orica claims "the downstream impact of variable and poorly controlled blast outcomes today can impact as much as 80 percent of the total mine processing costs" and adds that "this presents significant opportunity for the industry."[30] In other words, managing a blast can play a critical role in the productivity and profitability of a mine—the value mines derives from Orica's products and services. The better the mine manages the blast and controls the output, that is, the quality of the broken rock, the more money it can save. That is the niche Orica has defined for itself.

If Orica were to sell its explosives, advisory services, and other blasting materials under a traditional ownership model, or even under a subscription model, it would risk generating all forms of waste. The prices in a pay-to-own scheme might restrict access to the products and services by some mines, depending on the size of their upcoming jobs, the amount of explosives needed, and their financial wherewithal to buy what they need. Both pay-to-own and pay-per-time arrangements might generate consumption waste, unless customers can estimate with precision just how much product they need. Finally, these models might create performance waste if mines, for whatever reason, cannot generate quality blasts from the use of Orica's products. In this scenario, the financial consequences of a poor outcome rests solely on the shoulder of the mines, assuming of course that there are no defects in Orica's explosives themselves.

Orica reduced all three forms of waste by putting itself in the shoes of customers. For a success-dependent fee, Orica takes care of the necessary planning, provides the appropriate materials, and manages the blast. Just as important is Orica's steadfast focus on adopting new technologies to continually optimize how it integrates and performs these tasks. In late 2018, the company released the next generation of its digital platform, BlastIQ, which integrates data and insights from digitally connected technologies across the drill and blast process. Orica claims that the solutions powered by BlastIQ "can deliver predictable and sustainable improvements that can reduce the overall cost of drill and blast operations, improve productivity and safety, and facilitate regulatory compliance."[31] In other words, BlastIQ "will enable our customers to make better decisions, more rapidly and deliver improved blast outcomes across their operations."[32] It does that by providing "near real-time, hole by hole, blast-related data visually to the multiple users across the drill and blast process."[33]

Orica has reconfigured blast operations in mining around the idea of lean commerce. The adoption of Rock-on-Ground contracts creates opportunities for optimization that are inconceivable under the local, manual, highly variable ownership models that prevailed to that point. The difference is that the quality of the outcomes is now not only more quantifiable and verifiable, but also more predictable. Mining companies can make better decisions on how to conduct any given project, save a considerable amount of time and money, and capitalize on opportunities that might have otherwise been uneconomical without Orica's proposition.

This chapter and the previous two have shown how many companies in many industries around the world are using impact data, skills, experience, and creativity to shift their revenue models away from traditional, inefficient ownership schemes and toward agreements that come closer to reflecting the underlying value that customers derive in an exchange. Some of the models we described tackle access waste alone, while others aim higher by addressing consumption waste. Finally, the models discussed in this chapter attempt to quantify and track meaningful outcomes, if not value itself, and therefore focus on performance waste.

Part III of the book is all about action. Across five chapters, we will discuss the steps a company needs to take to develop and implement new revenue models successfully, and the challenges they are likely to face along the way.

III   Action

7 Committing to Outcomes

CNN Money once published a list of "America's biggest rip-offs." That eclectic collection of the most egregiously priced products and services included movie-theater popcorn, hotel mini-bars, and wine at restaurants. It also included college textbooks, whose prices allegedly reflect the wishes of "greedy publishers."[1] Between 1977 and 2015, prices of textbooks increased by 1,041 percent, or at three times the rate of inflation, according to an analysis of data from the United States Bureau of Labor Statistics.[2]

It is easy to scrutinize such lists, and the accompanying allegations of price gouging and profiteering, and ask several questions. In the case of textbooks, why are the prices so high? What should the "right" price be? What alternatives do students have, especially when second-hand books or library books may be hard to source?

In the spirit of the Ends Game, we feel that the solution to this particular "rip-off" should start with exposing and eliminating inefficiencies. Framed that way, the discussion leads to another set of interesting questions: how many students are denied access to educational materials because publishers insist on earning their revenue from the products themselves? How much waste occurs when students purchase textbooks but read only one or two chapters for a course? How much waste occurs when students feel that the textbooks they own contributed little to their learning objectives, be it a better course grade, the thrill of learning, or a specific job prospect? In short, do textbooks and other educational materials do the job that publishers intend them to do and teachers and students demand?

Knowing what truly drives educational outcomes—and what doesn't—would have far-reaching consequences for everyone who participates in education; from the school systems to their suppliers, and from teachers to students and their families. Inspired by such questions, United Kingdom–based Pearson—the self-described "world's learning company" and the leading publisher of educational materials—embarked on what it calls a "path to efficacy." In other words, Pearson decided to tackle the problem of high prices, and the intensive backlash to them, by first reviewing the very nature of the exchange it has with customers.

In 2012, Pearson published *The Learning Curve*, a report aimed at helping "policymakers, school leaders and academics identify the key factors that drive improved educational outcomes."[3] Then, in March 2013, two Pearson executive directors and one adviser published a report through the United Kingdom's Institute for Public Policy Research (IPPR) called *An Avalanche Is Coming: Higher Education and the Revolution Ahead* in which they described the "warped logic that has locked price and quality together" in higher education. "The price charged to students, even once the cost base is accounted for, is not always responsive to the classic relationship of supply and demand," they wrote. "Indeed, thanks to the inadequacy of outcome measures for universities … input measures tend to be seen as proxies for quality."[4] They argued that this logic has locked price together with a cost-based proxy for quality, and that this link "needs to be broken."

In 2013, Pearson made a full commitment to breaking this link when it decided to shift the focus of its business from selling educational materials to selling learning outcomes. It described the quest in literal terms in a regulatory filing with the U.S. Securities Exchange Commission. Pearson claimed its "path to efficacy" means it is "publicly committing to efficacy and improving learning outcomes. We will judge ourselves not by the products we make, but by their impact on learners. It will change how we decide which companies to acquire, where and how we invest, which products we get behind and which we retire. It changes how we recruit, train and reward each person in the company. This change will take time, and is why we talk about a 'path to efficacy' that we are on, and it is why we have committed to providing audited learning outcomes data for all our products and services by 2018."[5]

Since beginning the journey, Pearson has refined its aspiration as the pursuit of "greater impact on learner outcomes and learners' lives." Moreover, Pearson developed an "efficacy framework," which intends to focus the company's efforts on four key questions: What outcomes are we trying to achieve? What evidence do we have that we will achieve them? What plans and governance do we have in place to achieve them? And what capacity do we have to achieve them?[6]

In 2018, Pearson provided the first detailed reports and data on audited learning outcomes, under independent review by the auditing and accounting firm PwC. Pearson learned, for example, that its product suite for mathematics yielded only mixed results for students at universities. Overall, it noted that "increased attempts in quizzes and tests, increased average scores on quizzes and tests, and mastering a higher number of learning objectives were associated with statistically significant higher course grades."[7] Breaking the results down further, though, Pearson said: "Usage of MyLab Math and performance in quizzes and tests were significantly and positively associated with the two course outcomes: course grades and completion. However, use and performance on QuizMe was negatively associated with course grades." Pearson speculated that the discrepancy could be due to the fact that students work to reach a specific score threshold on QuizMe rather than to actually master the material.

Currently, Pearson charges students a yearly fee for access to MyLab materials, with the price and license period varying by subject matter and by the duration of a given course. That is, Pearson earns revenue on a pay-per-time scheme. The company collects impact data at the individual level from more than eleven million student users annually. The company claims that MyLab now "reacts to how students are actually performing, offering data-driven guidance that helps them better absorb course material and understand difficult concepts."[8] Importantly, as Pearson continues to gather impact data, it will gain sufficient evidence and confidence to make the basis of exchange with students some function of the improvements achieved. In other words, Pearson will have the opportunity to hold itself accountable for learning outcomes.

The Existential Question

The existential question for any company playing the Ends Game is: *What are we asking customers to pay for?* The truth about how organizations earn revenue lies in how they answer this question, and not in the promises made in advertisements, on websites, or face to face by sales people. Indeed, claims about offering superior value to customers are "cheap talk" unless organizations back up those claims by delivering the solutions customers seek to their needs and wants, *and* by adopting a revenue model that aligns its success with that of customers. The choice of revenue model is what drives accountability. If a company generates revenue on any basis other than the actual solutions it brings to the market, then that company is shortchanging customers—and ultimately itself—by possibility creating waste in the exchange.

Shifting to a relentless focus on outcomes—and preferably value itself—also makes an organization's innovation efforts more accountable. The curse of so many innovators is that they focus on making the product "better," only to realize that the tinkering has no link to actual customer benefits. If the product or service is superior to the offering of a competitor, but customers themselves are not interested this difference, then the innovator has wasted precious resources. But when the firm aligns its business goals with the satisfactions of customers via the choice of revenue model, the firm has the strongest possible incentive to focus research and development on the performance measure that the two parties have in common.

Pursuing this level of accountability would have been fantasy over a decade ago. The technological barriers to collecting and using impact

data were too high. But these barriers are disappearing, in some industries much faster than others, allowing (and at times pushing) organizations to finally "put their money where their mouth is." As we will
show in chapters 8–11, implementing a change from a less efficient
revenue model, probably one that is still based on the transfer of ownership of the product or service from the organization to customers, to
a revenue model based on desired outcomes is neither swift nor linear
nor certain. It presents a range of challenges, from the ability to collect
and analyze impact data without abusing the privilege, to ensuring that
customers are active and positive participants in the creation of quality
outcomes, and to the organization's own mindset, skills, commitment,
and resources to make the transition.

Defining "Outcome"

The starting point, clearly, is defining "outcome" in the organization.
From our perspective, there are four conditions that jointly determine
whether a given outcome is suitable as the basis for a revenue model.
First, the outcome must be meaningful—and therefore valuable—to
customers. This point is obvious, yet many businesses still fall into the
tempting trap of focusing on product or service attributes that they have
an inherent interest or competitive advantage in, yet these attributes
matter little to those who buy. Claims about meaningful outcomes—
which are the cornerstones of a firm's value proposition—could be objective or highly subjective, such as "enjoyment" in the case of Teatreneu.

Second, the outcome must be measurable using one or more parameters that are understood and accepted by the organization and its
customers. The organization must be able to quantify and express
its performance claims in a manner that can form the basis of the
exchange with customers. Customers must be able to verify the performance claims. Without these inputs, customers are exposed to possible
access, consumption, and performance waste. In business markets, for
example, perhaps the most basic outcome is that a particular product or
service improves the profitability of customers, either by lowering their

costs, increasing their revenue, or a combination of the two. But if profitability cannot be measured directly, then organizations must search for a parameter that can be observed. Orica adopted "broken rock" as a proxy for the impact of its offerings on mining operations.

Third, the measurement of the outcome must be robust, in the sense that the parameter is a faithful representation of the underlying outcome that interests the organization. Obviously, a low correlation between parameter and outcome challenges this requirement. In the worst cases, the correlation between the two could be zero or even negative, as Pearson discovered in the initial efficacy studies involving QuizMe. In the best cases, the company is able to define a functional relationship between parameter and outcome. Taking Pearson again as an example, its dedication to understanding efficacy ultimately led the organization to the point where it established a solid relationship between the use of MyLab by learners and outcomes such as course completion and course grades.

Finally, the measurement must be reliable, in the sense that neither customers nor a third party can tamper with it. That is, customers should not have the means to "fake" a performance level that is not accurate in order to derive a benefit. Some customers may try to outsmart a tracking system, especially when the measured variable has a direct effect on a price, a discount, or the opportunity to win a prize. One scenario is when customers attempt to underreport performance to reduce payment. The flipside is attempting to overreport performance to reach a specific reward, such as what happens with "fitness fraud." When people have a financial incentive to walk—such as a lower health insurance premium or discounts on fitness products—they can trick a step counter through any number of creative hacks, from metronomes and remote-control race cars to simply moving one's wrist while working or watching television.[9]

The most meaningful outcome, of course, is value itself. But measuring value can be difficult, despite all the technical improvements we described in chapter 2. For example, in consumer markets research, technology has not progressed to the point where a company can identify and measure (in a manner that is cost effective and scalable) the changes in brain activity that signal the pleasure individuals derive from everyday products and services. Even if this point could be reached, social norms may stop the company from collecting and using such intimate impressions, no matter how advantageous this could be for customers once a measure for "pleasure" is reflected in the revenue model. Accordingly, even in the most promising scenarios, firms typically have to select an outcome other than value itself that satisfies the four conditions above.

The Breadth and Depth of Outcomes

From a practical standpoint, outcomes have two primary dimensions: breadth and depth. The breadth of outcomes is a function of the heterogeneity of customer needs and wants. Conceptually, outcomes can vary by individual customer, by individual occasion, and across specific moments in time. Think of the automotive sector, where cars can serve a wide range of purposes. For example, some people seek the peace of mind of having a dedicated, 24/7 option to move around without delay or unwanted surprise. Others have specific transportation requirements (hauling equipment or animals, transporting children, etc.) that require a vehicle they can modify and customize as they see fit. A third segment of customers seeks flexibility because the objective changes constantly—commuting, business, grocery shopping, an off-road adventure, and so forth. A fourth group has a limited range of transportation alternatives, and therefore are "stuck" into a specific arrangement. This is especially true in rural areas. One manager at J.D. Power pointed out that the majority of Americans still live in a rural setting and need a car to get to work, a situation that means that "Uber is not an affordable alternative."[10]Finally, there are people who want a car simply because they derive satisfaction from knowing they own an exclusive object. It has nothing to do with actually using it!

On the other hand, the depth of outcomes is a function of complexity. Outcomes tend to be less complex when they depend only on the organization that develops the product or service and its customers, and when they can be broken down into a small set of clear, controllable steps. Conversely, outcomes tend to be more complex when they depend also on intermediaries and third parties, and when the underlying process is unclear or difficult to control. Ultimately, complexity is an issue of how many "moving parts" the organization has to keep track of and coordinate. The number of contributors is particularly important because, if a market evolves to the point where customers pay according to some measure of performance, then the "team" responsible for delivering that performance needs to agree on how to share the added value generated. For instance, this is likely to become an issue as autonomous vehicles grow in presence and acceptance. Transporting passengers or cargo efficiently from one point to another in an autonomous vehicle is clearly a complex outcome comprising the work of several stakeholders. The business that built the vehicle itself, the one that supplied its intelligence, the one that manages the onboard entertainment, and the one that coordinates traffic flows and navigation can all claim that they make a significant contribution to the quality of the outcome. The challenge facing these players is to settle on a revenue model that keeps waste to a minimum. We raise this as an exercise that requires thorough analysis and planning, not as an impediment.

The Quest vs. the Destination

Despite the emphasis of this chapter on the importance of embracing outcomes, the proper conclusion to draw is not that every organization can and should implement a performance model in the shortest possible time. For a variety of valid reasons, a company might not be able to reach the final destination of the Ends Game in the foreseeable future.

Instead, the proper conclusion to draw is that, today, revenue models anchored on the ownership of a product or service are patently inferior, and that making the transition to a revenue model anchored on time or use is certainly within reach of most businesses. What is important is to start the quest. A company needs to stretch and push itself to think in terms of outcomes, and thus focus on ends rather than means. If the jump turns out to be unrealistic or impractical, then the organization can still fall back on a revenue model that reduces access and consumption waste by following the lead of the companies we highlighted in chapter 5, or one that reduces only access waste, such as the examples featured in chapter 4.

"What we are confronted with in the drug industry is the existence of prices which by any test and under any standard are excessive."[1] That brief remark, which appeared in a report from a U.S. congressional subcommittee, will come as no surprise to anyone who pays attention to the cost of health care. The timing of that statement, however, may indeed come as a surprise. It was published in 1959.

Fast forward exactly sixty years and, once again, senior executives of seven leading pharmaceutical companies—this time AbbVie, Merck, AstraZeneca, Bristol-Myers Squibb, Janssen, Pfizer, and Sanofi—faced a U.S. Senate committee to answer questions about the high and rising cost of health care. The topics for that session sounded familiar. As National Public Radio pointed out, "high drug prices and profits, limited price transparency, aggressive marketing, alleged patent abuse and mediocre 'me too' drugs ... are identical to the issues senators investigated decades ago."[2]

One of the most provocative comments in the 2019 testimony came from Merck Chairman and CEO Kenneth Frazier, who told the committee: "The people who can least afford it are paying the most. That's the biggest problem we have as a country. We have a system where the poorest and the sickest are subsidizing others."[3] His comment echoed the sentiments of Heather Bresch, the CEO of the pharmaceutical company Mylan, who had testified before the U.S. House Oversight and Government Reform Committee in 2016 after her company's substantial price increases for a popular pharmaceutical product ignited a major controversy. Mylan had raised the list price for a two-pack of Epipen—an injector that can administer a potentially life-saving dose of epinephrine to someone in allergic shock—from $100 to more than $600 over the course of nine years, starting in 2007.[4] In a subsequent interview, Bresch shifted some of the blame for the high prices to other parts of the "system." She singled out insurers for setting high deductibles and drug-benefit managers for negotiating high discounts off of the high list prices.[5] "The irony is the system incentivizes high prices," a spokeswoman for Bresch said at the time.[6]

Given that politicians have grilled pharmaceutical executives regarding the same broken system for the last six decades, one might think that the problem of high drug prices is intractable. But is there a better way? In the spirit of the Ends Game, we rephrase this question and ask: how much inefficiency is there in the system? The answer is staggering. One study has estimated the amount of waste in the health-care sector in the United States at roughly $1 trillion, or 30 percent of the country's annual expenditure on health care.[7] Leaving aside high administrative costs and alleged outright fraud, a significant portion of this waste stems from differences in standards of care across states (which leads to access waste), overtreatment (which leads to consumption waste), and failures in care or coordination (which leads to performance waste).[8]

High prices for drugs seem to have no end in sight. In 2019, for example, Novartis announced that its gene-therapy treatment for an inherited disease called spinal muscular atrophy will cost a record-setting $2.125 million.[9] Given this outlook, how can medical systems address the inevitable waste that such high prices create? How can organizations, either individually or collectively, act in a way that makes them, their stakeholders (patients, insurance companies, other third parties), and society better off?

In our view, the path to answering this conundrum starts with the existential question *What are we asking customers to pay for?* Bill George, the former CEO of the medical technology giant Medtronic, hinted at one possible solution for the health-care sector in a commentary he wrote in response to Mylan CEO Bresch's congressional testimony in

2016. George argued that "authentic health-care companies from Mayo to Merck understand they are in business to restore people's health, and if they did that well, profits would follow."[10] In other words, success in the industry is the result of aligning financial performance with patient outcomes. In chapter 6, we described how some companies and institutions are already attempting to quantify patient well-being and make payments contingent on these measures. They are not alone. For example, another biotech firm, bluebird bio, plans to offer a gene-replacement therapy for an inherited blood disease on an installment plan tied to specific results. The company would get "as little as 20% of the product's total cost upfront, and put the remaining 80% 'at risk' of nonpayment."[11] Payments of an additional 20 percent would follow in the ensuing years, but only if the treatment meets certain criteria, such as eliminating or reducing the number of transfusions. "We only get paid if we do what we said we'd do," bluebird bio CEO Nick Leschly said.[12]

Nonetheless, many players within health-care systems continue to obsess over a different question. Instead of thinking hard about the merits of generating revenue by the pill, by the device, or by the treatment or procedure, these players think hard about the merits of a specific price point, and specifically about the opportunities to make products and services more affordable. Most other industries work the same way. It is as if the basic rules of the game in a given market, the nature of the revenue model, cannot be changed.

The Quality Paradox

Why don't organizations immediately leap at opportunities to play the Ends Game? Why doesn't an organization, knowing that eliminating waste unlocks market potential, act proactively to shake up the prevailing revenue model in its industry, trying to reach a better alignment with the value customers actually derive in an exchange?

All too often, the remarkable explanation is that such a company is "blinded" by the quality of the products and services it proudly brings to market. This is what we refer to as the *quality paradox*. At some point, the relentless pursuit of quality makes it almost unimaginable to generate revenue from anything other than the sale of one's offerings. Said differently, when a company obsessively directs its efforts toward continuously innovating its products and services, it risks becoming accountable to its *offering* rather than to its *customers*. In health care, for example, it implies that a pharmaceutical company, which spends considerable time, money, and effort developing new medications, focuses on pricing its own "outputs" rather than those that customers ultimately care about. Paradoxically, the better a company is at creating value for customers through its efforts at innovation, the more waste it generates in the exchange by continuing to adhere to the standard pay-to-own revenue model.

One probable cause of the quality paradox is *surrogation*, a concept made popular by Willie Choi, Gary Hecht, and William Tayler in a research article published in 2012.[13] Put in its simplest terms, surrogation occurs when an individual or institution becomes so keenly focused on improving the measure of an underlying construct of interest that it reaches a point where the measure replaces the construct entirely. Surrogation warps the intent behind the old management cliché about "what gets measured, gets done" into something like "what gets measured is everything." An often-cited example of surrogation is the 2018 credit card scandal involving Wells Fargo's retail banking: "Driven by strict and unrealistic sales goals, employees in Wells Fargo's Community Bank division engaged in fraudulent sales practices, including the opening of millions of fake deposit and credit card accounts without customers' knowledge."[14] What happens in such cases is that "a company can easily lose sight of its strategy and instead focus strictly on the metrics that are meant to *represent* it."[15]

Surrogation is likely to exist in the context of revenue models for two reasons. First, consider an organization that has a track record of innovation and invests constantly and heavily in research and development. This motivation may come from its desire to lead the market on quality and differentiation, or it may be a trait of the industry as a whole. Irrespective, the organization worships its products. However, similar to Gillette in chapter 4, and similar to many pharmaceutical or consumer electronics companies, such devotion reinforces attitudes and behaviors that are inherently inward-looking, making the organization less likely to consider—let alone accept—any metric for its revenue model other than the product or service itself. The deeper the roots in proprietary technologies or engineering and processes expertise, the more vulnerable is this organization to suffer from tunnel vision.

The second reason for surrogation in the context of revenue models is financial. Innovation is often an expensive exercise, and heavy investments tend to make organizations more conservative in any decision that involves revenue. Of the different models discussed in this book, the traditional ownership model is probably the simplest and "safest" to adopt, as it gives organizations a relatively clear sense of the costs involved in selling a product or service and, therefore, of the contribution they can expect from every transaction at a given price. This may not be the optimal contribution, but it is a less risky proposition.

In the spirit of the Ends Game, the creation of better outcomes for customers should go hand in hand with the creation of a better financial outcome for the organization. The tighter the alignment, the less waste occurs in the exchange between the two parties. If a company truly possesses a superior product or service, and especially when this company has the resources to innovate and maintain an advantage, then it does itself and its customers a disservice by stubbornly holding onto a revenue model based on ownership. An ownership model may prompt a dizzying price tag that reflects or validates the inherent quality of the offering. But this price tag may also deny access to potential customers and distort consumption patterns. Importantly, it offers no guarantee of performance.

In early 2019, the Boston Consulting Group cited an example of what would happen if a drug maker were to switch its revenue model for a cure from a per-treatment basis (a pay-to-own model) to a population-based payer licensing agreement analogous to the enterprise licensing model commonly used for software.[16] The firm used a cure for the hepatitis C virus (HCV) as its example. According to the data cited, most of the world's HCV-infected population remains uncured even though innovative medicines, introduced in 2013, can cure HCV in eight to twelve weeks.[17] The Boston Consulting Group's model of the HCV treatment showed that patients, payers, and pharmaceutical companies would do better under a payer licensing agreement agreement in every market they tested.[18]

As the Boston Consulting Group writes: "The advent of cures creates a true pricing dilemma for pharmaceutical companies and those that pay for medicines. If drugs were priced today to reflect only the value that accrues over time, the resulting (high) prices would strain payers, prompting some of them to limit patient access. Conversely, treating all patients as fast as possible—and in doing so, accelerating eradication—requires prices so low that developing certain cures would become far less economically attractive, particularly compared with the economics of drugs that treat chronic diseases."[19] In other words, amazing breakthrough cures do exists, but they cannot work their "magic" unless health-care systems adopt revenue models designed specifically to limit waste in the exchange between manufacturers and payers. The quality paradox hamstrings the consideration and adoption of such models.

Tackling the Problem

We could recommend at this juncture that companies in an industry with a broken "system" forge ahead by applying the concepts we introduced and elaborated in parts I and II of this book. They would take customer focus to its logical conclusion, using today's groundbreaking technologies to integrate impact data into their strategic calculus. They would seize growth opportunities in their markets, identifying where they can reduce the greatest amount of waste by aligning their revenue models with the desired outcomes customers achieve—or perhaps with value itself.

If only it were that easy. Before an organization can take these steps, it first needs to recognize that it faces the quality paradox. The organization needs to see that there is a tension between how it creates better outcomes for customers through its products and services (the solutions it brings to market) and how it creates better outcomes for itself (the revenue model). The stronger the solutions, the greater the tension if the organization blindly adheres to the same tried-and-true ownership model. Perhaps leaving aside industries such as luxury goods, where success is often closely tied to exclusivity, the organization must recognize that it makes no sense to improve an offering yet stick to a revenue model that blocks customers from purchasing, or one that dissuades customers from buying out of concern for inadequate consumption or performance.

Sticking to established revenue models leads to inertia, which can manifest itself as risk aversion, myopia, and a certain degree of hubris. The more companies are accustomed to perpetuating the past by projecting it into the future, the more likely they are to respond reactively to trends in their market instead of initiating them. An unwarranted commitment to an inferior revenue model is reinforced by efforts to win within the existing system—through cost cutting, process efficiency, and incremental innovation—instead of changing it. Recognition is the first step, because many managers are locked unnecessarily into a certain way of doing business.

The second step is to identify and appraise the opportunities and challenges that present themselves if the organization adopts a more efficient revenue model. The ease with which a company can make a change depends in part on the nature of the outcomes, as we described in chapter 7. For example, implementing a new revenue model grounded in complex outcomes is likely to require the participation of multiple players, which can be costly in terms of time, effort, and money. Having said that, companies that can consistently deliver best-in-class performance ultimately have the greatest incentive to play the Ends Game. Such organizations already focus their value proposition on superiority in one or more dimensions of quality, which suggests that they have a strong basis to implement a revenue model more closely aligned with the outcomes that result from this advantage. Depending on the dynamics of the market (the nature of the outcomes, the extent of competition, the existing level of transparency, etc.), implementing a revenue model that holds the firm accountable to its customers may in fact trigger a flight to quality. The irony, however, is that the companies most likely to benefit from implementing a better revenue model are those that are most likely to suffer from the quality paradox.

The impetus to make a change is unlikely to be strong among companies (or industries) that are currently performing well. This helps explain, for instance, the reluctance among many of the world's best pharmaceutical companies to move toward outcome-based revenue models. There is no doubt that the products and services these organizations bring to market already generate tremendous value for them, for patients, and for society at large. Columbia University professor Frank Lichtenberg analyzed the benefits of drugs in twenty-two countries and estimated that drugs launched since 1982 are adding 150 million life-years to people's life spans every year, at an average expenditure per life year of $2,837, an amount he considers to be a "bargain."[20] The pharmaceutical companies have clearly cashed in on this. One report from Bloomberg said that the top thirteen drug makers in the world earned over $100 billion in net income in 2018.[21]

However, a closer look tells a somewhat different story. The prices of many drugs are widely viewed as misaligned with the value they actually deliver. In a 2016 article appropriately titled "The Price-Quality Paradox in Health Care," the Health Care Cost Institute concluded: "The relationship between state-level quality and price measures demonstrate[s] how price alone may not be sufficient for identifying quality. In some cases, it appears that higher prices are actually associated with lower quality."[22]

The third and final step in overcoming the quality paradox is for the organization to persevere in search of the best possible answer to the question: *What are we asking customers to pay for?* Satisfaction with the current financial position should not distract the organization from this pursuit. As long as there is significant waste in the exchange with customers, there is room for improvement. More important, there is always the threat that the comfortable current position is "disrupted" by a bold competitor or new entrant that dares to look at the exchange and decides to offer customers a more efficient proposition.

We admit that answering this existential question is a grand, somewhat uncomfortable challenge. Frankly, it is tempting to skip the work and default to an answer that is familiar and "good enough." This remains one of the main reasons why so many companies resort to generating revenue directly off of the products and services they make. But innovative companies that deliver solutions to customers that are truly superior can no longer afford such complacency. These companies have the most to gain from resolving the quality paradox, but they are missing the conviction that they can be more successful, or sustain their success for longer, if they take customer focus full circle and rethink their revenue model from the perspective of those they are working hard to serve.

9 Getting Up Close and Personal

A clever discount scheme, ostensibly announced in 2018 by the Mexican flag carrier airline Aeromexico, took the idea of price promotions to an unprecedented and intimately personal level. Travelers from the southern United States could fly to Mexico for a price discounted by the percentage of Mexican DNA they had.

The two-minute video with the tagline "Innerdiscounts: There are no borders within us," showed surprised residents of the Texan town of Wharton learning that they could visit Mexico at, say, 15 percent or 18 percent off, after a DNA test revealed the extent of their Mexican ancestry.[1] This campaign offered the airline a way to educate potential customers about their particular connection to Mexico and perhaps encourage them to travel there and explore. The advertisement, released by the airline's ad agency Ogilvy[2] became a viral hit in early 2019 during the U.S. government shutdown, which was caused in part by differences on funding for a proposed border wall between the United States and Mexico.[3]

Apparently, however, the offer wasn't real.[4] The ad was "an experiment to see what would happen" and languished on YouTube with no budget until some self-promotion from the agency and the controversy around the border wall caused the video to go viral.[5]

Nonetheless, its existence raises important and delicate issues about data collection. How much information can and should organizations collect about their customers? What stake do customers have in making their data available? What guidelines should govern how the "gatherers" of data use and safeguard that information? In this chapter, we will look at what customers stand to gain and what they stand to lose when they willingly reveal details about themselves and their consumption behaviors. Understanding the stakes starts with realizing how these "impact data" differ from everything else organizations have been researching about customers in the past.

Three Types of Data

When we think about the information organizations may be interested in about their customers, we distinguish among three types. The first type helps companies understand what their customers truly need and want. This has been the lifeblood of marketing departments and research and development teams for decades. The seminal writings of Peter Drucker, Theodore Levitt, and several other scholars spurred a more intense and nuanced use of market research tools such as focus groups and quantitative surveys designed to get inside the customers' heads. Scientific and technological advancements have not only improved these methods, but also superseded them in specific cases. Today, approaches including direct observation and experiential interviews are increasingly popular due to their ability to tap into the subconscious thoughts and motivations of respondents.

The second type of data collected by organizations is relatively more recent and comprises information on the different steps that customers take to seek out and select solutions that supposedly satisfy their needs and wants. The original representations of these decision-making "journeys" were linear, with customers following a rather predicable path or "funnel" from awareness and interest to an actual purchase. However, the advent and growth of e-commerce both exposed and compounded the limitations of such a simple map of the purchase process. The decision journeys of twenty-first-century customers are anything but linear. They tend to unfold across multiple touchpoints and multiple channels, and organizations use this information to engineer rich experiences and forge stronger relationships with their target audience.

The third type of data is the impact data we defined in chapter 2. While collecting information on customers' needs, wants, and decision journeys is typically not an intrusive or invasive exercise, collecting impact data is by nature a deeply personal task, as they reveal what customers do with products and services after they gain access to them, as well as how well these offerings actually perform. The monitoring and tracking that underlies impact data can reveal facts that an individual or business customer purposely kept hidden. The data and subsequent analyses can also reveal patterns, tendencies, and behaviors that neither the organization nor its customers anticipated. For these reasons, most attempts to collect impact data require customers to opt in, on the assumption that the organization harvesting the information will protect it and use it primarily in the customers' interests.

Impact data enable organizations to take customer focus full circle and define a more efficient revenue model. Without impact data, in combination with traditional information on needs, wants, and journeys, there is no Ends Game because firms have no reliable means to identify and eliminate waste—they have no means to hold themselves truly accountable. Specifically, firms cannot switch to a revenue model that improves access and shifts the risks associated with consumption and performance from customers to themselves without knowing when and how its products are used, the context, extent, and effects of that use, and the resulting outcomes. The big question—probably the trillion-dollar question, if we look across all forms of access, consumption, and performance waste in an economy—is the extent to which customers are willing to share their information with firms and fuel the Ends Game.

Protecting Privacy and Building Trust

The bank robber was the folklore anti-hero in the nineteenth and twentieth centuries. Bank robber Willie Sutton purportedly said that he raided banks because "that's where the money is."[6] In the twenty-first century, the bank robber's counterpart is the hacker. According to one report, there were more than 3,800 data breaches in the first half of 2019.[7] In the health-care sector alone, there were 503 data breaches in 2018, three times more than in 2017.[8] Bank robberies in the United States, by the way, totaled a mere 2,975 incidences in 2018 according to the Federal Bureau of Investigation.[9]

The torch has been passed. The "money," in the form of consumption and performance data, is now stored on servers rather than in safes. In this context, any data-driven quest for a better revenue model may feel like theft to customers unless organizations take serious steps to protect their privacy and foster a sentiment of trust. Protecting privacy implies putting the appropriate safeguards in place to keep data confidential. This is a technological and regulatory challenge that lies beyond the scope of this book. Building trust, on the other hand, implies reassuring customers that the organization collects and uses impact data for purposes that ultimately are also in their interests.

A lack of trust is costly. In its thirteenth annual Global Consumer Pulse research, published in late 2017, Accenture stressed: "Poor personalization and lack of trust cost U.S. organizations $756 billion last year, as 41 percent of consumers switched companies. Without deeper customer insight, companies cannot deliver the experiences they crave."[10] Similarly, the *New York Times* underscored the importance of trust, and the unease about its absence, in an article describing the blatantly commercial motives of some companies that collected behavioral data at the individual level: "In recent years, data companies have harnessed new technology to immediately identify what people are watching on Internet-connected TVs, then using that information to send targeted advertisements to other devices in their homes."[11]

Finally, the software giant salesforce.com took this view one step further by using the term "crisis of trust." Importantly, the organization sees transparency and accountability as the basis for a solution, a way for customers to accept and even encourage personalization. In a 2018 report titled *The State of the Connected Consumer*, the company explained: "Delivering personalized experiences requires a data-driven, 360-degree view—but more than half of respondents are uncomfortable with how their data [are] used. Customers say companies can earn their trust by taking certain steps, such as giving them control over how their data are applied, and being transparent about how they are used. Eighty-six percent of customers are more likely to trust companies with their relevant information if they explain how it provides a better experience."[12]

Organizations face the trust challenge in a climate where customers are increasingly sensitive to the information that is gathered about them and their behaviors. In his book *TAP: Unlocking the Mobile Economy*, Anindya Ghose fittingly asks: Will the ability to collect and use personal data turn an organization into a concierge or a stalker?[13] Clearly, there is potential for a company to become a creepy stalker, especially seeing that data collection, driven by the breakthroughs in information technology we described in chapter 2, seems to have become more invasive and in some cases surreptitious.

Yet the reality is that sharing one's data has never really been a riskless proposition. Companies must be aware of the risk customers face and their sensitivity to it. They must do everything in their power to alleviate this concern. At the same time, however, companies must be able to communicate that sharing one's data has never been a more valuable investment for customers as it is today. The challenge for organizations playing the Ends Game, therefore, is to ensure that the balance between risk and reward ultimately favors the latter. It remains an educational and ethical challenge for organizations as they seek to adopt a revenue model that makes the exchange more efficient. Adopting a better revenue model eliminates waste, converting market potential into actual value. Colloquially speaking, this change grows the "pie" shared by the organization and its customers. Customers know that they play a critical role in making the pie bigger by allowing the organization to track consumption and performance. As such, customers deserve to share in the incremental benefit.

Accountability Is the New Bond

Lean commerce is shaped by important changes in the type of data that companies need, collect, and use. It is this information-dense "fuel" that powers the initiatives of the companies featured in this book including, for example, Winterhalter (chapter 5) and Orica (chapter 6). Successful organizations have immersed themselves fully into impact data in order to measure with the highest precision not only the consumption patterns of their discerning customers, but also the real benefits that they derive from products and services. The active and passive tracking of customers provides them with insights they could never have imagined ten or twenty years ago. Moreover, when customers know firsthand that an organization can use these data to deliver the outcomes they desire, it puts both parties in the exchange in an enviable position.

The critical ingredient that makes these relationships endure—the bond that keeps the data flowing, so to speak—is accountability. Customers demand that organizations live up to their promises. They accept to share their data with organizations because this will allow them to prove their worth. But one problem is that customers potentially generate streams of data that are more voluminous and intricate than a company can reasonably capture and process. Companies, therefore, must now sort the rapidly flowing and growing influx of data to find what it needs to create and, importantly, demonstrate value for their customers and themselves. Companies that succeed in channeling and filtering these streams of data seize the enormous opportunity to eliminate waste. They can encourage and even incentivize efficient consumption, improve the likelihood of more and better outcomes, and understand how they co-create value together with customers.

In business markets, the availability and use of impact data can lead to not only the measurement of outcomes, but also their prediction. For instance, the firm Syncron,[14] which claims to help companies "maximize product uptime," lists several valuable applications for data streams, such as predicting the remaining useful life of an aircraft engine, predicting the failure of electric submersible pumps used to extract crude in the oil and gas industry, and forecasting energy demand in small communities to predict the overload situations of energy grids.[15]

Consumer markets similarly offer opportunities for organizations to use customer data as a means to eliminate waste and enhance the customer experience. As we mentioned in chapter 8, reaching a destination via autonomous driving is a complex outcome. A flow of customer data is absolutely necessary to make it not only feasible, but also lucrative. A study by A.T. Kearney estimated that autonomous vehicles could save $1.3 trillion annually in the United States alone by reducing traffic accidents, cutting energy consumption, and lowering maintenance and service costs.[16] The operation of such vehicles depends on the interchange of massive amounts of data among passengers, vehicles, and central management and guidance systems. Autonomous vehicles also offer an important complex outcome: mobility. Put one way, the general goal of mobility is to "[s]eamlessly and intuitively assist passengers with where they are going, how they get there, and what they do along the way."[17] Accordingly, the exchange of data can have a direct influence on the revenue model, depending on how the provider of this complex outcome defines it. As the authors of the A.T. Kearney report write: "Driverless cars from Google are not only passenger transportation, but an ingenious data collection system. If passengers feel comfortable exchanging rich data—telemetry, pictures, and video—in exchange for a ride, Google can significantly lower the barriers to access. There is no reason, however, why the incumbent original equipment manufacturers cannot either pursue a similar strategy or neutralize Google's."[18]

Finally, the exchange of data can open up opportunities for organizations and customers to enhance their relationships even in something as seemingly simple as light bulbs. In an article on the Internet of Things (IoT), the Boston Consulting Group described the opportunity: "Smart bulbs transform consumers from occasional, anonymous buyers of light bulbs into consumers of connected lighting. If consumers opt in to the network, their usage data can yield insights that are valuable not only to the light bulb provider but also to other firms within the broader network, which could include utilities, interior designers, and consumer electronics firms. The challenge is to figure out how valuable the IoT data is to each of these potential customers."[19] The same thinking applies to large commercial installations as well. Signify (originally Philips Lighting) doesn't sell lamp installations (luminaires) to the Schiphol Airport in Amsterdam. It sells light, or more specifically, light as a service. One driving force for Signify is a commitment to a circular economy, which it describes as using "resources more effectively by creating rather than wasting, using rather than owning, and reusing rather than disposing."[20]

The most progressive organizations—especially startups that don't have to undo an entrenched, less efficient revenue model—promise customers something that most incumbents never seriously considered before: to profit only when customers do. As such, the more progressive organizations are collecting and exploiting new types of data to form a better picture of when customers use their offerings and how they perform. Success hinges on how well the company manages and capitalizes on this newfound transparency about its customers. This is where accountability and trust come into play, so that organizations and their customers work together to their mutual benefit.

10 Partnering with Customers

Are you a good driver?

Most people tend to answer this question with a resounding, confident "yes." This is not only our impression from personal experience, but also a well-documented finding in several academic studies.[1] Drivers seem particularly adept at overestimating their own skills. They consider that the greatest threat to their safety is not themselves, but "other drivers' actions that increase crash risk such as alcohol impairment or running red lights."[2] In fact, some drivers even think they have a better definition of what constitutes good driving than other people do.[3]

These self-serving beliefs can lead to frustration when people sign up for insurance. In the mind of the ubiquitous "above average" driver, the classic factors that determine an insurance premium, such as age, years of driving experience, place of residence, and accident history are clearly inadequate. To this driver, reality is more nuanced than a bunch of descriptors, and she in fact deserves a significantly lower rate than what a standard model would recommend. If only there were an insurance company that could see this problem and come up with a solution.

In the spirit of the Ends Game, the U.S.-based insurance provider Progressive took on the challenge and decided to offer customers the opportunity to adjust their insurance premiums to true, on-the-road driving behavior. To Progressive, "no two drivers are the same ... why should their insurance be?" Accordingly, the company uses "gamification" to encourage members to record and share their driving data in exchange for a personalized quote. In so doing, the company has transformed the murky process of setting the price of auto insurance, typically based on poor proxies of driving ability and risk pooling, into a fun, interactive contest that aligns the actions of customers every day and every mile with the price they finally pay.

The product that allows drivers to put their self-perceived reputations and skills to the test is called Snapshot, a pay-per-use insurance program that Progressive launched in 2008 to collect "information about how you drive, how much you drive, and when you drive." Snapshot gathers the necessary data via a smartphone app or a device that plugs directly into the vehicle (typically under the steering wheel). Specifically, it records the time of day the vehicle is driven, the number of miles driven, and sudden changes in speed (hard brakes and rapid accelerations).[4] A hard brake is defined as a decrease in speed greater than seven miles per hour in one second. Similarly, a rapid acceleration is defined as an increase in speed greater than seven miles per hour in one second. If the driver uses the smartphone app, Snapshot will also collect data on the driver's use of the handheld device while driving.[5]

Joining the Snapshot program begins with a thirty-day free trial with no obligation. Progressive sends drivers the plug-in device in the mail or asks them to download the Snapshot mobile app. Once a connection is initiated, Progressive starts tracking the required information and produces an estimate at the end of the trial period.[6] The majority of members see their rates go down after they join the program, with the average discount totaling $145.[7] But Snapshot is by no means a universal discount scheme. In 2015, Progressive reported that "eight of 10 Snapshot users get discounts, with the rest getting either no discount or a 'small' surcharge."[8] That same relationship held true in early 2019, when Progressive's website claimed that "only about 2 out of 10 drivers actually get an increase."[9] Safe drivers are more likely to self-select into a program such as Snapshot. Riskier drivers would prefer an alternative— either sold by Progressive or its competitors—that can effectively mask their true driving profile and allow safer drivers to subsidize them.

By 2019, Progressive had amassed a total of twenty-five billion miles of driving data, which it now uses not only to determine personalized rates for its customers, but also to gain a more complete picture of the risks involved in driving. For example, these data have allowed the company to confirm what until now had been an intuitive hypothesis, which is that drivers who "put their phones down are less likely to have a claim than those who drive with a mobile device in their hand."[10]

Stepping back, the example of Snapshot brings to light an important point. Organizations seldom determine outcomes and deliver value to customers solely on their own. That is, in many exchanges the creation of value is a shared responsibility of the organization *and* its customers. For Progressive, customers provide the necessary data that make pay-per-use insurance a reality *and* work together with the company to improve driving performance. Snapshot "gamifies driving, and discounts are earned over time as a way to encourage drivers to keep the app running, which requires having location tracking turned on."[11] What makes the difference is the extent of the discount, of course, but also the way Snapshot encourages customers to earn these rewards. The "Driving Details" feature allows members to "get an inside look at your driving, along with the tips you need to succeed at saving." The data and progress are updated daily, analogous to how some people may track their fitness, diet, or finances. The "Bragging Rights" feature rewards drivers with badges for improvement or for hitting milestones, and gives the participants a platform to "share how you're doing with friends and family, and challenge them to top your performance."[12] These game aspects close the cognitive gap within a group on how well each individual drives, and who deserves the title of best driver. The ability to post badges and rankings on social media is intended to enhance friendly competition and encourage participants to go that safer "extra mile" to improve performance and settle, once and for all, who is "above average" and who is not.

We mentioned earlier in part III that a firm may need to partner with other firms to realize a complex outcome. The issue now is that customers may also play an active role. Their contribution is likely critical. Yet, as organizations progress through the Ends Game, the risks associated with access, consumption, and performance shift from the shoulders of customers to those of the organization. If the revenue model supposedly renders customers indifferent to whether a product delivers outstanding value or not, then what motivates them to help generate quality outcomes?

The Challenge of Customer Indifference

When a revenue model is based on performance, the participation of customers is often essential. They have to act in a way that is consistent with the creation of quality outcomes. For example, a new health-care treatment may provide superior relief or even a cure, but it functions only if patients comply with the treatment plan. Unfortunately, compliance is often not a foregone conclusion. Aside from the outright refusal to take a given medication or undergo a given treatment, there is also the issue of unauthorized pill splitting, where patients feel they can tolerate the risk of subpar performance if this saves them money.[13] Similarly, the best machine in the world will not produce at peak performance if the user does not operate it, or does not operate it correctly. The best component cannot fulfill its potential if it sits in the warehouse or is not properly installed. As we discussed in chapter 6, Pearson provides an intrinsic incentive for learners to "do their part" in the creation of value because good grades or improved academic performance typically lead to better career prospects. But the tools offered by Pearson will not have the intended effect unless students use them and apply themselves.

If an organization needs customers to be active participants in the creation of value in the exchange, it must ensure that customers are not indifferent to the level of performance achieved. When customers are indifferent, they may display a casual disregard for their relationship with the organization. For example, subscription services such as Rent the Runway or any of the memberships offered by car manufacturers have a significant number of valuable assets in circulation. Their success—and even their reputation—depends on the ability to deliver a clean, functioning, attractive product to the next user, which in turn depends on the condition in which the previous user left the product. In that spirit, there is a "greater good" for customers to return dresses, vehicles, apartments, or any other shared items in good condition. Yet some customers may not perceive that they have that kind of implicit obligation.

Said differently, the less "skin" customers have in the game, the lower their motivation to play by the rules, and the greater their likelihood to engage in risky or exploitative behavior. In the spirit of the Ends Game, this means that the more risk one party bears in the success of an exchange, the less incentive the other party has to contribute to that success. This logic underpinned the work of Nobel Prize–winning economist Kenneth Arrow on moral hazard in the 1960s. In an interview in 2016, Arrow said that he felt that the U.S. health-care system "has not improved at all in terms of inefficiency" since he published his seminal paper in 1963.[14] "The system is about as inefficient, and it was pretty inefficient then too. The rather reasonable attempts to improve the delivery, that is to extend health care to more people, have led to a bigger system, and therefore more complexity and more chance for exploitation."[15]

As an organization adopts a revenue model that is more efficient, it also gradually adopts the risk associated with creating value for customers. Bearing this risk is not an issue if the organization is confident that, on its own, it can create quality outcomes consistently. When customers participate in this process, however, the quality of an outcome becomes increasingly less certain unless the company can offer customers the right incentive to make the proper contribution. Perhaps the simplest and most basic way to motivate customers is to make sure they benefit proportionally—if not disproportionally—as outcomes improve. That is, the more value customers help create together with the organization (that is, the more efficiency the two parties can drive in the exchange), the greater the share of that value they should retain.

Aside from financial incentives, organizations have three options to mitigate the risk they have assumed from customers: contracting, gamification, and integration.

Contracting

Today's technologies make it much easier to measure processes and outcomes with accuracy. To avoid ambiguity, the organization and its customers can take advantage of these technologies to define an outcome that is precise, measurable, and verifiable without the risk of manipulation.[16] The measurement of the outcome should also be independent of outside influences, or at least reconcilable with them.

But, in the course of the relationship, what happens if one side (e.g., customers) perceives that the actions or inactions of the other side (the organization) are doing harm more than they are doing good? Regardless of whether this judgment only reflects a subjective belief or is an objective reality, the doubting side might decide to cut back on its engagement. Oliver Hart, who won the 2016 Nobel Prize in Economics, and economic theorist John Moore refer to this phenomenon as "shading." Shading occurs "when a party isn't getting the outcome it expected from the deal and feels the other party is to blame or has not acted reasonably to mitigate the losses. The aggrieved party often cuts back on performance in subtle ways, sometimes even unconsciously, to compensate."[17]

To prevent or mitigate unexpected behaviors that threaten the achievement and integrity of an outcome, the organization and customers can enter into a formal "payment-by-results" contract. The contract exists to ensure that the two sides recognize their rights and obligations in the exchange. In particular, the organization wants customers to comply with the steps needed to deliver the highest level of performance possible. However, putting a contract in place presents several challenges, with reaching an agreement on measurability perhaps the hardest among them. The two biggest enemies of measurement in a payment-by-results contract are ambiguity and external influences.

If the contract, say, specifies the achievement of $500,000 in operational cost savings at a plant, that target may suffer from several influences that can have a material effect. These can include the loss of a key employee, the loss of a key account, or a change in the competitive environment, such as a new entrant or a merger. A customer might even neglect to comply with measurement protocols in an effort to pay less, perhaps because they conclude that "good enough" quality at a lower price is a better deal for them than "best possible" quality at a higher price. Therefore, the ability to measure performance both ex ante and ex post is a primary criterion for whether a payment-by-results contract makes sense.[18]

In our view, improving measurability involves three aspects: the organization needs a culture of measurement in place, it needs the technology to measure the outcome without ambiguity, and it needs the organizational capabilities to deliver the outcome within the agreed parameters. In terms of culture, the consulting firm Bain asserts that companies with a "measurement advantage" are "more likely to exceed their business goals, grow revenue and gain market share" than companies without it.[19] The key differences between the best and worst performers include a view of customer activity across all channels, access to real-time data, and a collaborative culture that allows the company to respond quickly to what it observes. The idea is that a company lacking a strong measurement culture would then struggle with the remaining aspects of sourcing the right technology and delivering the expected result.

Experience often plays an indispensable role in the decision to enter into a payments-by-results contract, defining the target outcome and deciding exactly how to measure it. A 2012 report featuring case studies from General Electric's (GE) Water and Power Division highlights the complexity and specificity that goes into measuring an outcome. One of many examples in the case of corrosion in aluminum cans at a brewery in Portugal. The GE unit first conducted an audit of parameters such as water quality, flow rates, water consumption, and biocide screening. The audit revealed a need for a change in biocide dosing and for a different corrosion inhibitor. Once these improvements were implemented, GE proceeded to fine-tune the program. The elimination of corrosion and the increased production of error-free cans saved the brewery $691,000.[20] Meanwhile, GE not only served a client successfully, but also underwent a learning experience that it can leverage in similar future situations to define plausible outcome ranges, manage expectations, establish contingencies, and better estimate the resources it might need to achieve the outcome desired by the client.

There is of course the possibility that a company in the position of GE could face cost overruns. Depending on the variance of previous outcomes, it would also need to account for underperformance or overperformance relative to the target. But these eventualities beg the question of whether a potentially risky contract is still superior (for both parties) over the long term than having the organization simply sell its products and services to customers and let them carry the responsibility for setting and achieving targets.

This can make sense, for example, when the organization is smaller and less powerful than the customer, which is relatively common in many business-to-business markets. The seller in this case might push for a payment-by-results contract, have it accepted, but then face a situation where the project exceeds targets and the customer refuses to pay what is owed and prefers to default to a different, more advantageous agreement. The organization faces what's called a hold-up problem. It could sue the customer for payment, but the process is uncertain, it is likely to take a long time, and it can seriously jeopardize the future of the relationship. Therefore, in a situation where customers are relatively powerful, the organization must be comfortable that the other party won't back out of the deal when it is convenient to do so. If the seller anticipates such a risk, it may opt for a conventional, more "arm's length" transaction where the transfer of ownership implies that the customer pays upfront.

Gamification

Gamification is a less intrusive option than creating obligations under a contract. The researcher Sebastian Deterding defines the practice as "the use of game design elements in non-game-contexts."[21] In turn, gamification is a common form of "nudging," a term made popular by Richard Thaler, yet another Nobel Prize–winning economist, and Cass Sunstein, a legal scholar and Harvard University Professor, that refers to "any aspect of the choice architecture that alters people's behavior in a particular way without forbidding any options or significantly changing their economic incentives."[22]

The two central components in a "game" are the set of rules or activities that engage customers and a clear rewards system. For a business, this implies combining customer tracking, data collection, and a set of goals and rewards into one attractive program. The data tracking combined with the badges, rewards, and bragging rights in Progressive's Snapshot program are a good example. Progressive states that the earning of a badge or the sharing of results has no direct bearing on the rate or premium customers receive. Yet the friendly competition—the nudge—indirectly leads to a better outcome because it encourages customers to both monitor and improve their metrics.

Mitsubishi Motors Corp. is taking that idea one step further through Road Assist+, an innovative smartphone app that "enables Mitsubishi drivers to enjoy the cost savings of consumption-based insurance without having to install telematics hardware devices in their vehicles or purchase a newer 'connected car' model."[23] Built in cooperation with LexisNexis Risk Solutions, Road Assist+ grants the control of telematics data to Mitsubishi, which it can then share with insurance companies.[24] It is the first car maker to implement such an approach, and to encourage people to sign up the company introduced a reward system that allows users to earn badges. The difference from Snapshot, though, is that Mitsubishi drivers can redeem the badges they earn in exchange for small rewards such as a gift cards and discounts on an oil change or car accessories.[25] Ultimately, Road Assist+ is meant to help accelerate customer acquisition and improve retention as the car maker tries to turn data into a new revenue stream.

Gamification can also apply when an outcome is hard to quantify, such as a positive experience, a feeling of comfort, or a feeling of excitement. The more a company can encourage paying customers to indulge or immerse themselves in a specific experience, the greater the value that both customers and the company receive. Two common examples of this idea are coffee breaks and sports. The popular Starbucks smartphone app uses gamification to enhance its bonus system, based on stars. Patrons typically accumulate two stars for every dollar they spend, but they can grow their account through bonuses, double-star days, or playing seasonal games.[26] The Boston Red Sox baseball team uses an even more elaborate points system to encourage season ticket holders to spend more money on games, events, and other activities related to the team. The ticket holders can even earn points by watching a game on television. They can redeem the points for perks such as autographed items, passes to go onto the playing field, passes to park in the same lot as the players, and messages on the scoreboard during the games. The season ticket holders compete to earn these rewards and can also use their points to bid on exclusive opportunities such as throwing out the first pitch or watching the game with the owners. Finally, the two members who earn the most average points during the season win a road trip to watch the Red Sox play in an opponent's ball park, a highly sought-after prize.[27]

Integration

In many markets, especially those for industrial products and services, superior technical knowledge and experience implies that the organization often has a better understanding of what it takes to achieve quality outcomes than customers do. Accordingly, one option is for the organization is to extend its operations and take over the activities that are typically undertaken by customers. In other words, the organization convinces customers to "step aside" and effectively outsource the work necessary to ensure the highest level of performance possible. Integration makes sense whenever customers lack the sufficient know-how, skills, or resources to ensure a result on par with what the organization could achieve on its own. It also makes sense as a last resort if the organization cannot motivate the customer sufficiently through incentives, contracts, or games. Taking this step clearly comes at a cost but makes sense when the organization believes that the upside is worth the investment or the opportunity cost from inaction are far greater.

Can the same kind of approach work in direct-to-consumer markets? The progress that Procter & Gamble's (P&G) new Tide Cleaners service has made since its launch in early 2019 may offer some insights.[28] P&G had experimented with subscription models for its market-leading Tide detergent brand ever since Dollar Shave Club and the other shave clubs successfully chipped away at the seemingly unassailable position of its Gillette subsidiary. As the *Wall Street Journal* pointed out in 2016, executives at P&G have focused "not only on what consumers buy but on how they buy" after being "blindsided" by the success of DSC.[29]

The website for this new service calls Tide Cleaners "the future of laundry care" and implies the value that comes from taking over some of the activities typically left to customers by stating "Tide Cleaners provides personalized washing, folding, dry cleaning, and alterations."[30] This raises the notion of delivering razor blades or Tide detergent pods to a new level. It would be invasive and awkward, to say the least, if Gillette or DSC sent a trained employee to a customer's residence to personally perform the act of shaving. But doing the laundry of customers—and thereby saving them from a workaround or a big expenditure—can qualify as a desirable and valuable experience. This is especially true, as one P&G executive pointed out, when "for many people, the closest laundry room is 20 floors down or 10 blocks down the street."[31]

Were P&G to change the revenue model to reflect the consumption of liquid detergent, powder, or pods, it would likely expose itself to the risk that customers manipulate or misreport usage to keep costs down. P&G has printed the estimated number of loads on its Tide containers for several years, but if customers know that the cost of the desired outcome (clean clothes) depends directly and proportionally on how much detergent they consume, they have an incentive to limit consumption. That is, by their actions, customers are likely to strike a balance between consumption and quality that is not in P&G's interest.

By actually *doing* the laundry for customers, the consumption of detergent no longer enters into the equation. P&G bears the cost entirely and now has its own incentive to achieve the performance expected by customers with the optimal use of resources. The ability to do laundry at massive scale can give P&G a competitive advantage in this market. Tide Cleaners also provides the organization with a considerable amount of impact data. As one analyst described it, Tide Cleaners gives the company "new, previously untapped data on the demographics and geographies of their most dedicated customers, it gives them an opportunity to test new products, and it gives them insight into [future] acquisitions."[32]

Firms should not expect or take for granted that customers automatically share their data and work effectively to achieve specific outcomes. A firm's commitment to accountability is expressed not only through its choice of revenue model, but also through the way it offers customers incentives to act as a partners.

The software pioneer Adobe became a global success by selling its offerings under the simplest of models: a plastic disc in a cardboard box. A user paid upfront for a perpetual license to use the company's software products. By early 2013, Adobe was generating $1 billion in revenue per quarter and earning a net profit margin of 19 percent. Fueled by consistent growth since the Great Recession, Adobe's stock price stood at a five-and-a-half-year high.[1]

Under those circumstances, few people would expect senior management to do much tinkering. After all, if the business ain't broke, why fix it? Rarer still would be to expect senior management to implement a bold plan likely to cause revenue growth to stall and profits to plummet in the short term.

Yet that is precisely what Adobe did. At its MAX conference in May 2013, the company dropped a bombshell. It would cease to sell a host of creative products—including Acrobat, Illustrator, InDesign, and Photoshop—on the basis of a perpetual license and, instead, offer them through a new subscription service it dubbed Creative Cloud. In other words, Adobe would make an immediate and full commitment to a revenue model focused on access rather the traditional one based on outright ownership.

"We believe we are at a key inflection point in the history of digital communications." This statement in Adobe's 10-K filing for the 2012 fiscal year hinted at the rationale behind the forthcoming change. David Wadhwani, Adobe's senior vice president and general manager of digital media, elaborated on the logic for the switch when he said that it would allow the organization to "put innovation in our members' hands at a much faster pace." Wadhwani added that, under the new subscription model, customers would receive enhancements on an ongoing basis, have access to the full range of Adobe's creative tools, and interact online with a community of peers.[2]

Adobe also recognized that such a bold move would not prosper if it left the market to its own devices. To make the benefits of Creative Cloud clear to potential customers, the organization undertook a broad promotional campaign. It set a goal to have face-to-face meetings with fifty thousand customers at events around the world to win them over. "We understand this is a big change, but we are so focused on the vision we shared for Creative Cloud, and we plan to focus all our new innovation on the Creative Cloud," explained Wadhwani, adding that "customer satisfaction for users using Creative Cloud has been off the charts."[3] Adobe also reached out to the investing community, stressing the need to be patient. The pitch to investors intended to make clear that Adobe was not adding uncertainty or instability unnecessarily. Rather, the organization was pushing a consistent stream of medium-term revenue into the future in exchange for tighter, longer-lasting relationships with customers. In fact, the more Adobe "lost" in the short term, the more it would earn in the future because it was replacing high, one-off revenue with a potentially perpetual stream of smaller monthly payments.

The abrupt switch to subscription, backed by that critical educational campaign with customers and investors, worked out as planned and ultimately helped Adobe reach new heights. By 2014, the anticipated short-term financial declines took hold, as revenues dropped by 6 percent and net profit by 67 percent from the peaks in 2012. But that period was followed by sustained improvements that would have been inconceivable had the organization clung to its historical pay-to-own model. Between 2014 and 2018, Adobe's revenue and profits increased at compound annual growth rates of 21 percent (to reach over $9 billion in absolute terms) and 76 percent, respectively. Even more impressive is the change in Adobe's market capitalization. In late June 2013, around one month after the MAX Conference at which Adobe announced the change to the subscription revenue model, its market capitalization stood at $22.5 billion. By September 2019, it increased by a factor of six to $134.5 billion.

When Innovation Is Wasteful

It seems almost trivial to claim that an organization benefits from introducing innovations that make customers better off. Yet the Ends Game challenges this statement, stressing that satisfying customers does not in itself guarantee success. Rather, an organization improves its chances of profiting from the innovations it brings to market as it adopts a revenue model that addresses the access, consumption, and performance problems that create waste in the exchange. The experience of Adobe and its shift to Creative Cloud supports this idea.

The successful, thirty-year-old software company from San Jose, California, had looked into the crystal ball and recognized that the traditional ownership model would soon become an impediment to its growth. Technological advances had exposed an important limitation of selling ownership in this sector: it is an inefficient, impractical means to make a steady stream of innovative products and services available to customers. Rather than trying to find a way to win within the status quo, Adobe imagined a different future, one in which it would hold itself more—but not fully—accountable for the promises it makes customers. It implemented a revenue model that essentially guarantees access to innovation.

Figure 11.1 illustrates the rationale behind Adobe's decision. The left line represents the value an organization delivers to customers via its products and services. The point on that line moves up as the organization introduces offerings that are increasingly meaningful to customers. On the other hand, the right line represents the revenue the organization generates from the exchanges with customers. The point on that second line moves up as the organization adopts a revenue model that is more efficient—that is, a revenue model that takes on

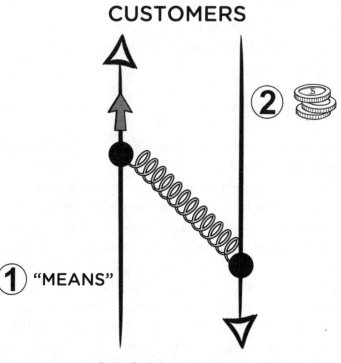

Figure 11.1
Organizations exist to (1) satisfy customers via their products and services and (2) earn revenue from them. Being closer to customers is great for the first task, but it creates inefficiency if the organization's revenue model does not follow suit.

increasing responsibility for possible access, consumption, and performance problems.

The spring connects the two points, and its tension captures the misalignment between the processes. The better a company understands what customers need and want, and the better it engineers their decision journeys, the more value it can theoretically deliver. Translating these insights into actual offerings and commercial actions moves the left part of the spring toward the top. However, if the revenue model does not make the organization accountable to its promise of superior value, the right part of the spring is "stuck" toward the bottom and waste is created. Adobe was in a position to launch such a stream of innovations in 2013, but it could not sell them effectively because the ownership model made them inaccessible.

An organization has two options to relieve the tension in the spring. First, it can cut back on its efforts to please customers—care less about understanding and satisfying their needs and wants or about engineering successful decision journeys. This is clearly a nonstarter. Second, the organization can revisit the revenue model, thinking about the waste created by forcing ownership, and changing the metric accordingly. Adobe ultimately made an aggressive move with objectives well aligned with what we have described in this book. The company believed that "market conditions presented significant opportunities ... to rapidly deliver product innovation, access new market segments, increase engagement with our customers, transition our business to promote a recurring revenue model, and accelerate our revenue growth."[4] Access models, thanks to the direct ongoing interaction with customers, can help companies introduce digital innovations faster, beat competitors to the market, and collect more and better feedback from customers sooner.

The misalignment and resulting tension shown in figure 11.1 are what Adobe saw when it looked into the crystal ball. The improvements offered by Creative Cloud, including the connectivity advantages and the flexibility conferred by cloud storage, created so much value for customers that the spring would have come under too much tension. Had Adobe insisted on sticking with the perpetual license model, it would have constrained its own ability to innovate and charge customers for it. It would have also made it virtually impossible for Adobe to bring further innovations to customers in a timely manner.

Adobe demonstrated that playing the Ends Game is worthwhile. But its decision to switch revenue models abruptly is not the only viable approach to relieving tension in the spring. Some companies decide to let old and new revenue models operate in tandem for some time in order to smooth the transition.

A Better Way to Buy Ink

Raise your hand if you have ever run out of ink or toner at work when you urgently needed to print a document. The relationship between a printer and its ink cartridges is structurally the same as the "razor and blades" model pioneered by Gillette we discussed in chapter 4.[5] Like the case of Gillette, the custom of selling ink cartridges on a pay-to-own basis creates access waste that leaves customers frustrated and, in the worst cases, without the ability to print what they need, when they need it. Buying ink cartridges historically has been a frustrating and inconvenient process, never mind an expensive one.

Unlike Gillette, however, HP seized the initiative and transformed its own business by launching a subscription program called Instant Ink. Under the Instant Ink program, the printer itself tracks ink levels and reorders cartridges when needed.[6] Monthly fees for the various plans are based on the number of pages the user prints per month. The printed pages could be anything the user desires, from black-and-white documents to color pages or photos downloaded from a phone. With respect to photos, HP hoped that easier and more convenient access to ink would help drive consumption. "We believe that many of the photographs taken today are in the digital jail," HP's current CEO Enrique Lores was quoted as saying. "You take many, many pictures, but you never take them out again. We want to release them."[7]

Instant Ink, which debuted in 2013, shares many elements with programs in other industries designed to broaden access. It has a value proposition that explicitly identifies a frustrating problem and offers a convenient solution. The user no longer needs to worry about when to restock ink, or worry about making emergency runs to an office supply store to pick up a cartridge. Critically, the metric underlying the revenue model is no longer ownership, but time. In fact, because subscribers can carry over unprinted pages from month to month, Instant Ink effectively morphs into a consumption model where the metric is the number of pages printed rather than monthly access.

According to one report, Instant Ink attracted two million subscribers by the end of 2017.[8] On the company's earnings call with analysts in February 2019, Dion Weisler, the then president and CEO of HP, said that the Instant Ink subscriber base "continues to have impressive growth. We have now rolled out the program in 18 countries and continue to see strong adoption rates."[9] But HP, unlike Adobe, did not make a wholesale, sudden switch to Instant Ink. Rather, customers who needed ink could still purchase it the old-fashioned way by visiting a local retailer or going online.

Generally speaking, operating multiple revenue models at the same time raises two important issues. First, when given a choice, existing customers will switch to whatever revenue model makes them better off. This will probably improve retention and ultimately steal share from competitors, but it also implies that the short-term impact on sales to existing customers is negative, which in turn can create friction inside an organization that is caught off guard. For example, although HP decided to maintain the traditional "pay-per-cartridge" model as it introduced Instant Ink, what happens if the subscription program prompts a significant cannibalization of existing sales? This can create animosity between teams, with the established unit responsible for cartridges perhaps doubting the logic of a new revenue model if it implies "shooting oneself in the foot." Internal politics clearly matter and must be managed, keeping in mind both the short-term consequences of giving customers options and the desired long-term benefits of making the transition to a more efficient revenue model. Equally important is the careful calibration of the prices used under each revenue model, so that cannibalization is not rampant.

Second, assuming that two or more revenue models should not coexist indefinitely because they result in different levels of efficiency, what is the actual transition plan? If the organization thinks only of the efficiency gains that come from aligning the metric by which it earns revenue with the way customers derive value, then the goal is to transition away from ownership as quickly as the size of the opportunity demands. But the organization must also account for the different costs of making this transition—including the actual costs of retiring the old revenue model and introducing the new one, the likely opportunity costs, and so on. The nature of the products (physical vs. digital), the pace of technological change in the market, and beliefs about how long it will take customers to change habits are likely to be important factors in this calculus. Consistent with this idea, HP faces a much different environment than Adobe does. Adobe's product is a digital one, which made the direct and complete switch to Creative Cloud a reasonable move. But HP sells a tangible, physical good. Adobe also faced a much faster pace of technological change than HP does.

The Many Faces of Internal Resistance

Any organization, regardless of its position in the market, ultimately needs to examine the gap between the value it creates for customers and the way in which it earns revenue for itself. The opportunity to eliminate waste by loosening the tension in the "spring" and gradually bringing the revenue model in line with the outcomes customers desire is universal. However, when it comes to rethinking the revenue model, companies typically face considerable internal resistance—a dangerous mix of inertia, neglect, myopia, and fear of change. The challenge is to identify the root causes, motivate senior leadership to act, and then guide the organization through the transition.

Sir Isaac Newton defined inertia in his first law of motion, when he stated that "every object will remain at rest or on its current trajectory unless compelled to change its state by the action of an external force."[10] If we wanted to convert this statement into Newton's first law of strategy, then we would need to make two substantial changes: Every *company* will remain at rest or on its current trajectory unless compelled to change its state by the actions of external *and internal* forces. In commerce, the external force that often overcomes inertia is technology, which in the context of revenue models is creating the unprecedented opportunities we first described in chapter 2. But this remains little more than untapped potential unless the organization makes a conscious decision to address the inefficiency that currently exists in the exchange with customers, no matter how safe or comfortable the status quo may feel.

Similarly, neglect or outright myopia may also block an organization from taking swift action. These factors can lead a company to focus on improving an existing but inferior process rather than striving to rethink the process altogether. As Amazon CEO Jeff Bezos explained in his 2016 letter to shareholders: "As companies get larger and more complex, there's a tendency to manage to proxies. This comes in many shapes and sizes, and it's dangerous, subtle." This comment relates to the notion of surrogation we cited in chapter 9 to explain the roots of the quality paradox. But Bezos goes a step further than merely talking about how metrics may grow to supersede the underlying goals they are meant to measure, and describes how entire processes can become all-consuming: "If you're not watchful, the process can become the thing. The process becomes the proxy for the result you want. You stop looking at outcomes and just make sure you're doing the process right."[11]

Finally, common business clichés provide convenient comebacks when people are confronted with the possibility of making a change as major as a shift in the revenue model. They will worry about the risks, because the step appears too much of a departure from "the way we've always done things." For example, the prospect of cannibalization is a recurring argument against acting. Why should the company give customers an opportunity to pay less? Isn't this a case of that familiar expression "the cure is worse than the sickness"? Especially in the age of viral social media, managers in an organization may worry excessively about how customers will respond and how the changes might affect the investing community. Yet when an organization feels comfortable with processes and patterns of behavior, these are hard to challenge and harder still to change. The resistance is stronger for incumbents, which carry the burden of past decisions, than it is for startups, which are free from any legacy.

Getting It Right

There are many reasons why established organizations are reluctant or hesitant to summon that internal "force" needed to change their revenue model—a move tantamount to changing the very nature of how they make a living in the market. However, as we said in the introduction, accountability is no longer a fashionable marketing slogan. It is a strategic imperative, and an organization's choice of revenue model determines the extent to which it holds itself to account for the promises it makes to customers.

In our mind, this choice is important and needs to be inspired by the observation that the exchange between organization and customers is at its most efficient when the incentives between the two parties are aligned—that is, when the organization benefits on the same dimension that defines value for customers. Any revenue model that departs from this ideal generates some waste, but this of course may be reasonable if the cost of further efficiency gains outweighs the benefit.

Accordingly, the best place to start for an organization is by reviewing the existing revenue model. Under the current regime, what are the biggest sources of waste in the exchange with customers—access, consumption, or performance? While answering this question, keep in mind that these three sources of waste are interrelated. Indeed, earlier we referred to them as natural checkpoints toward the solutions or outcomes that customers desire. Specifically, while customers clearly cannot derive value from the offerings they find in the market unless these perform as expected, performance is in itself contingent on consumption (customers cannot perceive performance in something they don't even use), and consumption is in turn contingent on access (customers cannot consume or experience something they don't even have).

The sequential nature of this relationship is important because it should motivate the organization to aim high. For instance, the shift to access models is a popular course of action today, and it is conceivable that the first move away from ownership eliminates the bulk of the inefficiency in the relationship between organization and customers.

Adobe's bet on Creative Cloud is a good example. But the so-called membership economy comprises only the first stage in the Ends Game, and taking on the responsibility for a higher-order form of waste such as consumption or even performance implies also taking on the responsibility for the waste that may come before it. There is a clear incentive for the organization to continue evolving.

Without much delay, however, the organization also needs to ask itself what business would look like if it dealt with customers on the basis of outcomes. Imagining this scenario requires creativity and the proper perspective. What is the right benchmark when a firm judges a future course of action? Although this is often the case in practice, the point of comparison should not be the status quo—as expressed by current performance in terms of the key financial and commercial indicators. This confers a false sense of security. Rather, the organization should draw a comparison between multiple futures, contrasting the likely consequence of a change in revenue model with the likely consequence of inaction (that is, the decision to maintain the existing revenue model). Moreover, the proper time horizon for this comparison is not the short term. As Adobe demonstrated, a change in revenue model can have negative immediate repercussions as the customer base "adjusts" to the new conditions. In fact, it helps to view any immediate dip on sensitive metrics such as number of customers, revenue, or profitability as an investment in a more sustainable future.

Any company that prides itself on being close to customers and their needs and wants will need to revisit its revenue model at some point if it wants to take this strategy full circle. This happens only when the company takes a step back from its daily routine, pushes back the pressures that come from internal processes, and starts questioning the gap between what it promises customers and what they actually pay for. What type of exchange do we want to be held accountable for moving forward? Similarly, what type of exchange will our customers hold us accountable for in the future?

Notes

1 Unfinished Business

1. D. Thompson, "The History of Sears Predicts Nearly Everything Amazon Is Doing," *The Atlantic*, September 25, 2017, https://www.theatlantic.com/business/archive/2017/09/sears-predicts-amazon/540888/.
2. Thompson, "The History of Sears Predicts Nearly Everything Amazon Is Doing."
3. A. Kharpal, "Alibaba Breaks Singles Day Record with More Than $38 Billion in Sales," CNBC, November 11, 2019, https://www.cnbc.com/2019/11/11/alibaba-singles-day-2019-record-sales-on-biggest-shopping-day.html; see also M. Kaplan, "Alibaba's 2019 Singles Day: $38 Billion; 200,000 Brands; 78 Countries," Practical Ecommerce, November 14, 2019, https://www.practicalecommerce.com/alibabas-2019-singles-day-38-4-billion-200000-brands-78-countries.
4. A. Selyukh, "Long Kept Secret, Amazon Says Number of Prime Customers Topped 100 Million," NPR, *The Two Way* (blog), April 18, 2018, https://www.npr.org/sections/thetwo-way/2018/04/18/603750056/long-kept-secret-amazon-says-number-of-prime-customers-topped-100-million.
5. Amazon, "Alexa, How Was Prime Day? Prime Day 2019 Surpassed Black Friday and Cyber Monday Combined," press release, July 17, 2019, https://press.aboutamazon.com/news-releases/news-release-details/alexa-how-was-prime-day-prime-day-2019-surpassed-black-friday.
6. A. Smith, *An Inquiry into the Nature and Causes of the Wealth of Nations* (Chicago: University of Chicago Press, 2008), Kindle, 6.
7. Smith, *An Inquiry into the Nature and Causes of the Wealth of Nations*, 6.
8. "Henry Ford: The Man Who Taught America How to Drive," *Entrepreneur*, October 8, 2008, https://www.entrepreneur.com/article/197524.
9. "Henry Ford."
10. P. Vlaskovits, "Henry Ford, Innovation, and That 'Faster Horse' Quote," *Harvard Business Review*, August 29, 2011, https://hbr.org/2011/08/henry-ford-never-said-the-fast.

11. "Guru: Alfred Sloan," *The Economist*, January 30, 2009, https://www
.economist.com/news/2009/01/30/alfred-sloan.

12. K. Marr, "Toyota Passes GM as World's Largest Automaker," *Washington Post*, January 22, 2009, http://www.washingtonpost.com/wp-dyn/content/article/2009/01/21/AR2009012101216.html?noredirect=on.

13. "Annual Model Change Was the Result of Affluence, Technology, Advertising: Roaring Twenties Made It Clear That People Were Buying Status and Novelty, Not Just a Ride," *Automotive News*, September 14, 2008, http://www.autonews.com/article/20080914/OEM02/309149950/annual-model-change-was-the-result-of-affluence-technology-advertising.

14. "Annual Model Change Was the Result of Affluence, Technology, Advertising."

15. V. Howard, "The Rise and Fall of Sears," *Smithsonian Magazine*, July 25, 2017, https://www.smithsonianmag.com/history/rise-and-fall-sears-180964181/.

16. Z. Meyer, "Timeline: Sears' Rise and Fall as Nation's Top Retailer," *USA Today*, March 22, 2017, https://www.usatoday.com/story/money/business/2017/03/22/sears-timeline/99488226/.

17. "Creating Mass Markets: Mass Distribution," from *Railroads: The Transformation of Capitalism* exhibition and catalogue, Harvard Business School Historical Collections, 2012, https://www.library.hbs.edu/hc/railroads/mass-distribution.html.

18. Howard, "The Rise and Fall of Sears."

19. E. McClelland, "Sears Is Dying: What the Ubiquitous Store's Death Says about America," *Salon*, April 13, 2014.

20. L. H. Jenks, "Railroads as an Economic Force in American Development," *The Journal of Economic History* 4, no. 1 (May 1944): 1–20.

21. M. McLuhan, *Understanding Media: The Extensions of Man* (Berkeley, CA: Gingko Press, [1964] 2002), Kindle, 113.

22. C. Sandburg, "Chicago" in *Poetry* (Poetry Foundation, 1914), https://www.poetryfoundation.org/poetrymagazine/poems/12840/chicago.

23. City of Chicago, "Chicago History," https://www.cityofchicago.org/city/en/about/history.html, accessed June 5, 2019.

24. "U.S. Ton-Miles of Freight," *National Transportation Statistics, 2018 4th Quarter* (U.S. Department of Transportation, Bureau of Transportation Statistics, 2018), https://www.bts.gov/content/us-ton-miles-freight, accessed April 13, 2020.

25. McLuhan, *Understanding Media*, 352.

26. S. Mintz and S. McNeil, "The Rise of Mass Communication," *Digital History*, 2018, http://www.digitalhistory.uh.edu/disp_textbook.cfm?smtID=2&psid=3315, accessed November 18, 2019.

27. J. L. Cruikshank and A. W. Schultz, *The Man Who Sold America: The Amazing (but True!) Story of Albert D. Lasker and the Creation of the Advertising Century* (Boston: Harvard Business Review Press, 2010), Kindle, 247.

28. C. Hopkins, *My Life in Advertising and Scientific Advertising* (New York: McGraw-Hill Education, Advertising Age Classics Library, 1966), Kindle, 224.

29. Hopkins, *My Life in Advertising and Scientific Advertising*, 229.

30. "Gallup Research," *AdAge*, September 15, 2003, https://adage.com/article/adage-encyclopedia/gallup-research/98482/.

31. Gallup, "Recognized as One of the World's Most Influential Americans. George H. Gallup, Founder | 1901–1984," https://www.gallup.com/corporate/178136/george-gallup.aspx, accessed June 5, 2019.

32. P. F. Drucker, *Management: Tasks, Responsibilities, Practices* (New York: HarperCollins, 1993), Kindle, 64.

33. R. Reeves, *Reality in Advertising* (New York: Widener Classic, [1961] 2015).

34. Amazon, "About Amazon," https://www.aboutamazon.com/?utm_source=gateway&utm_medium=footer, accessed February 8, 2020.

35. IKEA, "Vision & Business Idea," https://www.ikea.com/gb/en/this-is-ikea/about-the-ikea-group/vision-and-business-idea/, accessed June 5, 2019.

36. D. Ogilvy, *The Unpublished David Ogilvy* (London: Profile Books, 2012), 87.

37. "John Wanamaker," from *Who Made America?*, PBS, https://www.pbs.org/wgbh/theymadeamerica/whomade/wanamaker_hi.html, accessed June 5, 2019.

38. J. Surowiecki, "The Priceline Paradox," *New Yorker*, May 22, 2000, 34.

2 Beyond Needs and Journeys

1. PhRMA, "Biopharmaceutical Research & Development: The Process behind New Medicines," 2015, http://phrma-docs.phrma.org/sites/default/files/pdf/rd_brochure_022307.pdf.

2. U.S. Food and Drug Administration, "FDA Approves Pill with Sensor That Digitally Tracks if Patients Have Ingested Their Medication," news release, November 13, 2017, https://www.fda.gov/newsevents/newsroom/pressannouncements/ucm584933.htm.

3. U.S. Food and Drug Administration, "FDA Approves Pill with Sensor."

4. Volkswagon UK, "Warranty," https://www.volkswagen.co.uk/owners/warranty.com, accessed October 16, 2019.

5. F. Reichheld, "The One Number You Need to Grow," *Harvard Business Review*, December 2003, https://hbr.org/2003/12/the-one-number-you-need-to-grow.

6. Gartner, "Gartner Identifies Top 10 Strategic IoT Technologies and Trends," press release, November 7, 2018, https://www.gartner.com/en/newsroom/press

-releases/2018-11-07-gartner-identifies-top-10-strategic-iot-technologies-and -trends.

7. J. Aguilar, "'Data Is the New Asphalt': High-tech Colorado Road Test to Be First of Its Kind in the U.S., May Improve Traffic and Save Lives," *Denver Post*, May 30, 2018, https://www.denverpost.com/2018/05/30/us-285-smart-pavement -technology/.

8. J. B. Snyder, "New Tesla Features Make Car Sharing Easier," *Autoblog*, August 21, 2017, https://www.autoblog.com/2017/08/21/new-tesla-model-3-features-car sharing-smartphone-app-key-card/.

9. A. Jardine, ""Thanks 2016, It's Been Weird," Says Spotify in Biggest-Ever Global Campaign," *AdAge*, November 28, 2016, https://adage.com/creativity/ work/thanks-2016/50063.

10. United States Security and Exchanges Commission, Spotify Form 20-F filing, 2018, https://www.sec.gov/Archives/edgar/data/1639920/000156459019002688/ ck0001639920-20f_20181231.htm.

11. Capgemini, "Marketers Must Leverage the Power of Insight-driven Personal-ization and Use a Predictive or Prescriptive Approach to Understand the Needs and Desires of Customers," *Creating a Segment of One*, June 20, 2018, https://www .capgemini.com/2018/06/creating-a-segment-of-one/.

12. A. Galkin, "Retail Switch: From Generalization to Hyper-Personalization," *Forbes*, June 25, 2018, https://www.forbes.com/sites/forbestechcouncil/2018/06/ 25/retail-switch-from-generalization-to-hyper-personalization/#3a0682736bc0.

13. Gallup, "Recognized as One of the World's Most Influential Americans. George H. Gallup, Founder | 1901–1984," https://www.gallup.com/corporate/ 178136/george-gallup.aspx, accessed October 16, 2019.

14. B. Harper, "A Look at Ford's Vision of the Future: Cars Connected to Every-thing," *Driving*, August 16, 2018, https://driving.ca/ford/auto-news/news/a -look-at-fords-vision-of-the-future-cars-connected-to-everything.

15. Ericsson, "Digital Transformation and the Connected Car," *Next Stop: Smarter Cars*, November 2016, https://www.ericsson.com/en/mobility-report/digital -transformation-and-the-connected-car.

3   Leaner Commerce

1. N. Koukakis, "Shocking Truth: 20% of Health-care Expenditures Wasted in US and Other Nations," CNBC, January 13, 2007, https://www.cnbc.com/2017/ 01/12/shocking-truth-20-of-health-care-expenditures-wasted-in-us-and-other -nations.html; Business Wire, "Amazon, Berkshire Hathaway and JPMorgan Chase & Co. to Partner on U.S. Employee Healthcare," news release, January 30,

2018, https://www.businesswire.com/news/home/20180130005676/en/Amazon
-Berkshire-Hathaway-JPMorgan-Chase-partner-U.S.

2. Business Wire, "Amazon, Berkshire Hathaway and JPMorgan Chase & Co. to Partner on U.S. Employee Healthcare," news release, January 30, 2018, https:// www.businesswire.com/news/home/20180130005676/en/Amazon-Berkshire -Hathaway-JPMorgan-Chase-partner-U.S.

3. INRIX, "Los Angeles Tops INRIX Global Congestion Ranking," news release, February 5, 2018, http://inrix.com/press-releases/scorecard-2017/.

4. D. Morris, "Today's Cars Are Parked 95% of the Time," *Fortune*, March 13, 2016, http://fortune.com/2016/03/13/cars-parked-95-percent-of-time/.

5. A. Long, "Urban Parking as Economic Solution," International Parking Institute (December 2013), 42–45, https://www.parking.org/wp-content/uploads/ 2016/01/TPP-2013-12-Urban-Parking-as-Economic-Solution.pdf.

6. T. Higgins, "The End of Car Ownership," *Wall Street Journal*, June 20, 2017, https://www.wsj.com/articles/the-end-of-car-ownership-1498011001.

7. L. Johnson, "Digital Advertising Is Facing Its Ultimate Moment of Truth and Billions of Dollars Are at Stake," *Adweek*, September 4, 2017, http://www .adweek.com/digital/digital-advertising-is-facing-its-ultimate-moment-of-truth -and-billions-of-dollars-are-at-stake/.

8. U.S. Department of Education, "The Federal Role in Education," May 25, 2017, https://www2.ed.gov/about/overview/fed/role.html.

9. National Public Radio, "Can More Money Fix America's Schools," *All Things Considered*, April 25, 2016, https://www.npr.org/sections/ed/2016/04/25/ 468157856/can-more-money-fix-americas-schools.

10. B. Caplan, *The Case against Education: Why the Education System Is a Waste of Time and Money* (Princeton, NJ: Princeton University Press, 2018), back cover flap.

11. T. Levitt, as quoted in A. Gallo, "A Refresher on Marketing Myopia," *Harvard Business Review*, August 22, 2016, https://hbr.org/2016/08/a-refresher-on -marketing-myopia.

12. P. F. Drucker, *Managing for Results* (New York: HarperCollins e-books, 2009), Kindle, 94.

13. OECD Health Division, *Tackling Wasteful Spending on Health* (Paris: OECD Publishing, 2017), https://www.oecd.org/els/health-systems/Tackling-Wasteful -Spending-on-Health-Highlights-revised.pdf.

14. L. S. Dafny and T. H. Lee, "Health Care Needs Real Competition," *Harvard Business Review*, December 2016, https://hbr.org/2016/12/health-care-needs-real -competition; emphasis added.

15. J. Clawson, P. Lawyer, C. Schweizer, and S. Larsson, "Competing on Outcomes: Winning Strategies for Value-Based Health Care," Boston Consulting

Group, January 16, 2014, https://www.bcg.com/publications/2014/health-care
-payers-providers-biopharma-competing-on-outcomes-winning-strategies-for
-value-based-health-care.aspx.

16. Clawson et al., "Competing on Outcomes."

17. W. Buffett and J. Bezos, as quoted in Business Wire, "Amazon, Berkshire
Hathaway and JPMorgan Chase & Co. to Partner on U.S. Employee Health-
care," news release, January 30, 2018, https://www.businesswire.com/news/
home/20180130005676/en/Amazon-Berkshire-Hathaway-JPMorgan-Chase
-partner-U.S., accessed April 13, 2020.

4   Shaving, Rocking Out, and Looking Fabulous

1. A. Sorkin and S. Lohr, "Procter Said to Reach a Deal to Buy Gillette in $55
Billion Accord," *New York Times*, January 28, 2005, https://www.nytimes.com/
2005/01/28/business/procter-said-to-reach-a-deal-to-buy-gillette-in-55-billion
-accord.html; N. Deogun, C. Forelle, D. K. Berman, and E. Nelson, "P&G to Buy
Gillette for $54 Billion," *Wall Street Journal*, January 28, 2005, https://www.wsj
.com/articles/SB110687225259838788.

2. S. Terlep, "Gillette, Bleeding Market Share, Cuts Prices of Razors," *Wall Street
Journal*, April 4, 2007, https://www.wsj.com/articles/gillette-bleeding-market-share
-cuts-prices-of-razors-1491303601.

3. T. Heath, "How Hipster Brands Have the King of Razors on the Run," *Washington Post*, April 5, 2017, https://www.washingtonpost.com/business/capital-
business/how-hipster-brands-have-the-king-of-razors-on-the-run/2017/04/05/
edca3af6-1a27-11e7-9887-1a5314b56a08_story.html?utm_term=.0d1c2e6e6ffd.

4. C. Isidore, "P&G to Buy Gillette for $57B," *CNN Money*, January 28, 2005,
https://money.cnn.com/2005/01/28/news/fortune500/pg_gillette/.

5. B. Jopson, "Procter & Gamble: Time to Freshen Up," *Financial Times*, July 28,
2013, https://www.ft.com/content/e1782cc4-e95a-11e2-9f11-00144feabdc0.

6. J. Trop, "How Dollar Shave Club's Founder Built a $1 Billion Company That
Changed the Industry," *Entrepreneur*, March 28, 2017, https://www.entrepreneur
.com/article/290539.

7. J. Neff, "Why P&G's $57 Billion Bet on Gillette Hasn't Paid Off Big—Yet,"
AdAge, February 15, 2010, https://adage.com/article/news/marketing-p-g-s-57
-billion-bet-gillette-years/142116/.

8. Neff, "Why P&G's $57 Billion Bet on Gillette Hasn't Paid Off Big."

9. J. Trop, "How Dollar Shave Club's Founder Built a $1 Billion Company that
Changed the Industry," *Entrepreneur*, March 28, 2017, https://www.entrepreneur
.com/article/290539.

10. B. Booth, "What Happens When a Business Built on Simplicity Gets Complicated? Dollar Shave Club's Founder Michael Dubin Found Out," CNBC, March 24, 2019, https://www.cnbc.com/2019/03/23/dollar-shaves -dubin-admits-a-business-built-on-simplicity-can-get-complicated.html; M. Isaac and M. J. de la Merced, "Dollar Shave Club Sells to Unilever for $1 Billion," *New York Times*, July 20, 2016, https://www.nytimes.com/2016/07/20/ business/dealbook/unilever-dollar-shave-club.html.

11. Terlep, "Gillette, Bleeding Market Share, Cuts Prices of Razors."

12. Harry's, "Our Story," https://www.harrys.com/en/us/our-story, accessed April 13, 2020.

13. J. Koblin, "Investors Bet Big on Harry's, The Warby Parker of Shaving," *New York Times*, December 3, 2014, https://www.nytimes.com/2014/12/04/style/ investors-bet-big-on-harrys-the-warby-parker-of-shaving.html.

14. Trish Christoffersen, "Our Common Core: Create Fun and a Little Weirdness," Zappo's, https://www.zappos.com/about/core-values-three, accessed October 21, 2019.

15. A. Massoudi, "Unilever Buys Dollar Shave Club for $1bn," *Financial Times*, July 20, 2016, https://www.ft.com/content/bd07237e-4e45-11e6-8172-e39ecd3b86fc.

16. S. Terlep, "Dollar Shave Club's $1 Billion Deal: A Victory for Simplicity over Technology," *Wall Street Journal*, July 20, 2016, https://www.wsj.com/ articles/dollar-shave-clubs-1-billion-deal-a-victory-for-simplicity-over-technol ogy-1469044731.

17. P. Sisson, "Self-Storage: How Warehouses for Personal Junk Became a $38 Billion Industry," *Curbed*, March 27, 2018, https://www.curbed.com/2018/3/27/ 17168088/cheap-storage-warehouse-self-storage-real-estate.

18. H. Simon, *Confessions of the Pricing Man* (New York: Springer, 2006), 91–92.

19. B. Atwood, "The History of the Music Industry's First-Ever Digital Single for Sale, 20 Years after Its Release, *Billboard*, September 13, 2007, https://www .billboard.com/articles/business/7964771/history-music-industry-first-ever -digital-single-20-years-later.

20. Goldman Sachs, "Music in the Air: Streaming Drives Industry Comeback," *Insights*, December 12, 2006, https://www.goldmansachs.com/insights/pages/ music-in-the-air.html.

21. A. Newcomb, "Spotify Trounces Apple Music in Competition for Streaming Music Service Paid Subscribers," *Fortune*, February 6, 2019, http://fortune.com/ 2019/02/06/spotify-apple-music-paid-subscribers-streaming-music-service/.

22. Reuters, "Global Recorded Music Industry Grew Nearly 10 Percent Last Year: IFPI Report," news release, April 2, 2019, https://www.reuters.com/article/

us-music-growth-report/global-recorded-music-industry-grew-nearly-10
-percent-last-year-ifpi-report-idUSKCN1RE1WP.

23. Warner Music Group, "Warner Music Group Reports Results for Fiscal Fourth Quarter and Full Year Ended September 30, 2018," news release, December 20, 2018, http://www.wmg.com/news/warner-music-group-corp-reports-results -fiscal-fourth-quarter-and-full-year-ended-september-3-5.

24. M. Singleton, "Apple Announces New Apple Music App, Replacing iTunes, at WWDC 2019," *Billboard*, June 3, 2019, https://www.billboard.com/articles/ business/8514115/apple-announces-new-music-app-replacing-itunes-wwdc -2019.

25. L. Thomas, "Gillette One Ups Dollar Shave Club with On-Demand Razor Ordering Service Where You Text to Order," CNBC, May 9, 2017, https://www .cnbc.com/2017/05/09/gillette-one-ups-dollar-shave-club-with-on-demand -razor-ordering-service-where-you-text-to-order.html.

26. Gillette, https://gillette.com/, accessed February 8, 2020.

27. Gillette, https://www.gillette.co.uk/blog/subscriptions/how-gillettes-subs cription-work/, accessed February 8, 2020.

28. M. Isaac and de la Merced, "Dollar Shave Club Sells to Unilever for $1 Billion."

29. Dollar Shave Club, https://www.dollarshaveclub.com/, accessed June 6, 2019.

30. M. Malamut, "Why Are Razors So Darn Expensive?," *Boston Magazine*, August 9, 2013, https://www.bostonmagazine.com/health/2013/08/09/why-are -razors-so-expensive/.

31. J. M. Izaret, N. Hunke, J. Pineda, F. Fabbri, and W. Chia, "Cloudified Pricing— Coming to an Industry Near You," Boston Consulting Group, May 24, 2018, https://www.bcg.com/publications/2018/cloudified-pricing-coming-industry -near-you.aspx.

32. J. Hagel, J. S. Brown, M. Wooll, and A. de Maar, "Align Price with Use," Deloitte, February 13, 2016, https://www2.deloitte.com/insights/us/en/focus/ disruptive-strategy-patterns-case-studies/disruptive-strategy-usage-based -pricing.html#endnote-sup-3.

33. AWS, "Slack Case Study," https://aws.amazon.com/solutions/case-studies/ slack/, accessed April 13, 2020.

34. M. Helft, "Netflix to Deliver Movies to the PC," *New York Times*, January 16, 2007, https://www.nytimes.com/2007/01/16/technology/16netflix.html.

35. AWS, "Netflix Case Study," https://aws.amazon.com/solutions/case-studies/ netflix/, accessed June 6, 2019.

36. Rent the Runway, https://www.renttherunway.com/, accessed April 13, 2020.

37. S. Maheshwari, "Rent the Runway Now Valued at $1 Billion with New Funding," *New York Times*, March 21, 2019, https://www.nytimes.com/2019/03/ 21/business/rent-the-runway-unicorn.html.

38. Rent the Runway, "Pick a Plan," https://www.renttherunway.com/plans ?act_type=top_nav_anon_gated_t3, accessed February 8, 2020.

39. M. Colias, "GM Tries a Subscription Plan for Cadillacs—A Netflix for Cars at $1500 per month," *Wall Street Journal*, March 19, 2017, https://www.wsj.com/ articles/gm-tries-a-subscription-plan-for-cadillacsa-netflix-for-cars-at-1-500-a -month-1489928401?mod=article_inline.

40. "What Are Car Subscription Services?," Edmunds, January 8, 2019, https:// www.edmunds.com/car-leasing/what-are-car-subscription-services.html.

41. B. Howard, "GM Halts Pricey Book by Cadillac Car Swap Subscription Plan," Extremetech.com, November 5, 2018, https://www.extremetech.com/ extreme/280107-gm-halts-pricey-book-by-cadillac-car-swap-subscription-plan.

42. M. Wayland, "Cadillac's Revived Subscription Program to Rely on Dealers," *Automotive News*, January 20, 2019, https://www.autonews.com/dealers/ cadillacs-revived-subscription-program-rely-dealers.

43. Pernod Ricard, "Pernod Ricard's Breakthrough Innovation Group Unveils Opn at CES," news release, January 4, 2017, https://www.pernod-ricard.com/en/ media/press-releases/pernod-ricards-breakthrough-innovation-group-unveils -opn-ces-0/.

44. Iryna, "Pernod Ricard: Digitizing the World of Premium Drinks," *Digital Innovation and Transformation* (blog), Harvard Business School, April 22, 2018, https://digit.hbs.org/submission/pernod-ricard-digitalizing-the-world-of -premium-drinks/.

5 Flying Hours, Wash Cycles, and Miles Driven

1. D. Waldstein, "A Bloody Finger Sidelines Trevor Bauer Early, But the Indians Barely Flinch," *New York Times*, October 18, 2006, https://www.nytimes.com/ 2016/10/18/sports/baseball/cleveland-indians-toronto-blue-jays-trevor-bauer -alcs.html.

2. D. Sheinin, "A Man and His Drone: Indians' Pitcher Trevor Bauer Marshals Eccentricities," *Washington Post*, October 16, 2016, https://www.washing tonpost.com/sports/nationals/a-man-and-his-drone-indians-pitcher-trevor -bauer-marshals-eccentricities/2016/10/16/35f6e430-93de-11e6-bc79 -af1cd3d2984b_story.html?utm_term=.a5453ac41b57.

3. J. Spero and D. Bond, "Can Airports Ever Make Themselves Safe from Drones?," *Financial Times*, January 10, 2019, https://www.ft.com/content/ df78e80c-1413-11e9-a581-4ff78404524e.

4. "UK Insurtech Flock Launches Pay-As-You-Fly Drone Insurance," *Insurance Journal*, March 13, 2018, https://www.insurancejournal.com/news/international/ 2018/03/13/483092.htm.

5. "Huge Volumes of Data Make Real-time Insurance a Possibility," *Economist*, September 21, 2017, https://www.economist.com/finance-and-economics/2017/09/21/huge-volumes-of-data-make-real-time-insurance-a-possibility.

6. E. L. Klinger, "How to Insure a Flying Robot," LinkedIn, October 25, 2017, https://www.linkedin.com/pulse/how-insure-flying-robot-ed-leon-klinger/.

7. Flock, "Simpler, Smarter Drone Insurance," https://flockcover.com/ accessed April 13, 2020.

8. Flock, "Flock's Data Insights Part One: Pilots Really Do Fly Safer with Flock," October 15, 2018, https://blog.flockcover.com/flocks-data-insights-part-one-pilots -really-do-fly-safer-with-flock-40391adb7c26.

9. Flock, "Introducing: Recreational Drone Insurance for the Price of a Coffee," May 28, 2018, https://blog.flockcover.com/introducing-recreational-drone -insurance-for-the-price-of-a-coffee-47530dc4720f.

10. "Huge Volumes of Data Make Real-time Insurance a Possibility," *Economist*, September 21, 2017, https://www.economist.com/finance-and-economics/ 2017/09/21/huge-volumes-of-data-make-real-time-insurance-a-possibility.

11. "Huge Volumes of Data Make Real-Time Insurance a Possibility."

12. J. Willcox, "Cord Cutting Continues, Fueled by High Cable Pricing, Consumers Reports' Surveys Finds," *Consumer Reports*, September 17, 2019, https:// www.consumerreports.org/phone-tv-internet-bundles/people-still-dont-like -their-cable-companies-telecom-survey/.

13. R. Molla, "The History of Netflix Price Increases in a Single Chart," *Vox*, January 16, 2019, https://www.vox.com/2019/1/16/18185174/netflix-price-increase -subscription-chart-original-content-streaming.

14. YouTube TV and HBO, sign-up pages, https://tv.youtube.com/welcome/ and https://www.hbo.com/ways-to-get, prices effective as of February 8, 2020.

15. A. Pressman, "How Cord Cutting Is Driving Big Changes Across the Media Landscape," *Fortune*, June 5, 2019, http://fortune.com/2019/06/05/cord-cutting -netflix-apple-pwc/.

16. J. Fox, "How to Succeed in Business by Bundling—and Unbundling," *Harvard Business Review*, June 24, 2014, https://hbr.org/2014/06/how-to-succeed-in -business-by-bundling-and-unbundling.

17. Apple, "iTunes Store Sets New Record with 25 Billion Songs Sold," news release, February 6, 2013, https://www.apple.com/newsroom/2013/02/06iTunes -Store-Sets-New-Record-with-25-Billion-Songs-Sold/.

18. C. Kitchener, "Doing Dishes Is the Worst," *Atlantic*, April 3, 2018, https:// www.theatlantic.com/family/archive/2018/04/doing-dishes-is-the-worst/ 557087/.

19. Winterhalter Company Presentation 2017, Prezi, updated July 24, 2019, https://prezi.com/6ku90zjxjqre/winterhalter-company-presentation-2017/.

20. "No Investment, Zero Risk, Pay Per Wash," Winterhalter, https://www.pay -per-wash.biz/int_en/, accessed April 13, 2020.

21. "No Investment, Zero Risk, Pay Per Wash," Winterhalter UK, http://www .pay-per-wash.biz/uk_en/, accessed April 13, 2020.

22. H. Simon, F. Bilstein, and F. Luby, *Manage for Profit. Not for Market Share* (Cambridge, MA: Harvard Business School Publishing, 2006), 51–52.

23. Winterhalter Company, "70 Years of Winterhalter," news release, FCSI, August 20, 2017, https://www.fcsi.org/industry/products/70-years-winterhalter/.

24. Winterhalter Company Presentation 2017.

25. G. Crossan, S. Hupfer, J. Loucks, and G. Srinivasan, "Accelerating Agility with Everything-as-a-Service," Deloitte, September 17, 2018, https://www2 .deloitte.com/insights/us/en/industry/telecommunications/everything-as-a -service-xaas-flexible-consumption-models.html.

26. Crossan et al., "Accelerating Agility with Everything-as-a-Service."

27. R. Wilson, "HE Payne Opt for Michelin EFFITIRES™ Programme," Commercial Tyre Business, September 11, 2018, https://www.commercialtyrebusiness .com/latest-news/posts/2018/september/he-payne-signs-up-to-michelin -effitires-policy/.

28. Michelin, "Michelin Solutions Go-Ahead with EFFITIRES™ Extension," news release, September 27, 2016, https://news.cision.com/michelin-solutions/ r/michelin-solutions-go-ahead-with-effitires--extension,c2087888.

29. Michelin, Michelin America's Truck Tires, https://www.michelintruck.com/, accessed April 13, 2020.

30. World Economic Forum, "Michelin Solutions," *Digital Transformation*, http://reports.weforum.org/digital-transformation/michelin-solutions/?doing_ wp_cron=1557721029.2728700637817382812500.

31. Michelin, "Services and Solutions," https://www.michelin.com/en/activities/ related-services/services-and-solutions/, accessed April 13, 2020.

32. "Michelin Acquires a Telematics Company, and 7,000 Fleet Customers," *Modern Tire Dealer*, June 14, 2017, https://www.moderntiredealer.com/news/ 722889/michelin-acquires-a-telematics-company-and-7-000-fleet-customers.

33. B. Straight, "From a Tire Company to an Innovative Solutions Provider, Michelin Grows Its Legacy," FreightWaves, June 1, 2018, https://www .freightwaves.com/news/technology/michelin-is-no-longer-just-a-tire-company.

34. K. Wong, "How to Become a Digital Nomad," *New York Times*, February 27, 2019 https://www.nytimes.com/2019/02/27/travel/how-to-become-a-digital -nomad.html.

35. H. Meyer and B. Katz, "At Ziferblat, It's Pay by the Minute," *Bloomberg Businessweek*, May 14, 2015, https://www.bloomberg.com/news/articles/2015-05-14/ ziferblat-moscow-pay-by-the-minute-cafe-opens-franchises-abroad.

36. S. Bains, "Pay-per-minute Café Ziferblat Offering Unlimited Free Food and Drink Is Coming to Birmingham," *Birmingham Live*, July 16, 2018, https://www.birminghammail.co.uk/whats-on/food-drink-news/pay-per-minute-cafe-ziferblat-14896165.

37. Snakes and Lattes, "Contact," https://www.snakesandlattes.com/contact/#about, accessed April 13, 2020.

38. Snakes and Lattes, "Contact"; L. Ipsum, "Snakes & Lattes Is Opening a Third Board Game Café in Toronto," *Daily Hive*, April 5, 2017, https://dailyhive.com/toronto/snakes-lattes-board-game-cafe-midtown-toronto-2017.

39. C. Lieber, "The Share Economy Is Coming for Your Closet," *Vox*, October 16, 2018, https://www.vox.com/the-goods/2018/10/16/17983756/fashion-rentals-rent-the-runway-tulerie-armarium.

40. Tulerie, https://www.tulerie.com/, accessed June 6, 2019.

41. SpotHero, "Frequently Asked Questions," https://spothero.com/faq/, accessed June 6, 2019.

42. TaskRabbit, https://www.taskrabbit.com/, accessed June 6, 2019.

43. Trringo, https://www.trringo.com/, accessed June 6, 2019.

44. PwC, "Europe's Five Key Sharing Economy Sectors Could Deliver €570 Billion by 2025," news release, 2016, https://www.pwc.com/hu/en/pressroom/2016/sharing_economy_europe.html.

45. Z. Yiran, "Sunny Days Ahead for Sharing Economy Firms as Transaction Volumes Rise," *China Daily*, March 15, 2019, http://global.chinadaily.com.cn/a/201903/15/WS5c8b1018a3106c65c34eece0.html.

46. Yiran, "Sunny Days Ahead for Sharing Economy Firms as Transactions Volumes Rise."

47. M. Cliffe, "European 'Sharing Economy' Tipped for Rapid Growth," ING, https://www.ingwb.com/insights/articles/european-sharing-economy-tipped-for-rapid-growth, accessed April 13, 2020.

6 Laughter, Rocks, and Quality of Life

1. "Glassworks Project 'Pay Per Laugh' Wins 8 Lions at Cannes," *Little Black Book*, 2014, https://lbbonline.com/news/glassworks-project-pay-per-laugh-wins-8-lions-at-cannes/; "Teatreneu/Pay Per Laugh," Clio Awards, 2015, https://clios.com/awards/winner/public-relations/teatreneu/pay-per-laugh-1214.

2. "Teatreneu—'Pay Per Laugh,'" Adforum, https://www.adforum.com/creative-work/ad/player/34498880/pay-per-laugh/teatreneu, accessed April 13, 2020.

3. D. Basulto, "An Innovative New Payment Model That's No Laughing Matter," *Washington Post*, October 14, 2014, https://www.washingtonpost.com/

news/innovations/wp/2014/10/14/an-innovative-new-payment-model-thats
-no-laughing-matter/.

4. "Fact Sheet: Audiences," Shakespeare's Globe Theatre, https://teach
.shakespearesglobe.com/fact-sheet-audiences; L. Clark, "Evidence of a Seating
Plan Discovered at the Colosseum," *Smithsonian*, January 26, 2015, https://www
.smithsonianmag.com/smart-news/please-find-your-seats-evidence-seating
-plan-discovered-colosseum-180954023/.

5. E. Çano, R. Coppola, E. Gargiulo, M. Marengo, and M. Morisio, "Mood-Based
On-Car Music Recommendations," *Lecture Notes of the Institute for Computer Sciences* 188 (2016): 154–163.

6. J. White, "A Car That Takes Your Pulse," *Wall Street Journal*, November 28,
2012, https://www.wsj.com/articles/SB10001424127887324352004578131083
891595840.

7. K. Fitchard, "Forget Apps, Ford's OpenXC Project Will Produce Open-Source
Car Hardware," *GigaOm* (blog), January 10, 2013, https://gigaom.com/2013/01/10/
forget-apps-fords-openxc-project-will-produce-open-source-car-hardware/.

8. G. Peoples, "The Growing Cost of Music's Monthly $9.99 Price Tag," *Bill-
board*, September 20, 2019, https://www.billboard.com/articles/business/8530616/
music-streaming-prices-competition-subscribers.

9. "Driver Health Monitoring," *MIT Technology Review*, March 1, 2018, https://
insights.techreview.com/7-breakthrough-car-technologies-to-watch-driver
-health-monitoring/.

10. K. Fitchard, "How Gracenote Is Building a Car Stereo That Senses Your
Driving Mood," *GigaOm*, February 19, 2013, https://gigaom.com/2013/02/19/how
-gracenote-is-building-a-car-stereo-that-senses-your-driving-mood/.

11. G. Topham, "The End of Road Rage? A Car Which Detects Emotions,"
Guardian, January 23, 2018, https://www.theguardian.com/business/2018/jan/
23/a-car-which-detects-emotions-how-driving-one-made-us-feel.

12. W. Oremus, "Google's Big Break," *Slate.com*, October 13, 2013, https://slate
.com/business/2013/10/googles-big-break-how-bill-gross-goto-com-inspired
-the-adwords-business-model.html.

13. L. Flynn, "With Goto.com's Search Engine, the Highest Bidder Shall Be
Ranked First," *New York Times*, March 16, 1998, https://www.nytimes.com/1998/
03/16/business/with-gotocom-s-search-engine-the-highest-bidder-shall-be-ranked
-first.html.

14. Oremus, "Google's Big Break."

15. Google Ads, https://ads.google.com/home/, accessed April 13, 2020.

16. Alphabet Inc., Form 10-K, *Annual Report 2018*, United States Security
and Exchange Commission, https://www.sec.gov/Archives/edgar/data/1652044/
000165204419000004/goog10-kq42018.htm.

17. Alphabet Inc., Form 10-K, *Annual Report 2018*.

18. J. Greenwood, "The Future of Drug Pricing: Value over Volume," *The Hill*, October 11, 2017, https://thehill-com.cdn.ampproject.org/c/s/thehill.com/opinion/healthcare/354913-the-future-of-drug-pricing-value-over-volume?amp.

19. National Institute for Health and Care Excellence glossary, https://www.nice.org.uk/glossary?letter=q, accessed April 13, 2020.

20. D. Roland, "Obscure Model Puts a Price on Good Health—and Drives Down Drug Costs," *Wall Street Journal*, November 4, 2019, https://www.wsj.com/articles/obscure-model-puts-a-price-on-good-healthand-drives-down-drug-costs-11572885123.

21. J. Whalen, "What Is a QALY?," *Wall Street Journal*, December 1, 2015, https://www.wsj.com/articles/what-is-a-qaly-1449007700.

22. Roland, "Obscure Model Puts a Price on Good Health—and Drives Down Drug Costs."

23. A. Frappé, "The Call for Further Advancement of Indication-based Pricing," *PharmExec.com*, February 1, 2018, http://www.pharmexec.com/call-further-advancement-indication-based-pricing.

24. Roche, "Innovative Pricing Solutions," https://www.roche.com/sustainability/access-to-healthcare/innovative-pricing-solutions.htm, accessed April 13, 2020.

25. A. Pollack, "Pricing Pills by the Results," *New York Times*, July 14, 2007, https://www.nytimes.com/2007/07/14/business/14drugprice.html.

26. Amgen Corporation, "Amgen and Harvard Pilgrim Agree to First Cardiovascular Outcomes-Based Refund Contract for Repatha® (Evolocumab)," news release, May 2, 2017, https://www.amgen.com/media/news-releases/2017/05/amgen-and-harvard-pilgrim-agree-to-first-cardiovascular-outcomesbased-refund-contract-for-repatha-evolocumab/; Repatha, https://www.repatha.com/.

27. Amgen Corporation, "Amgen Makes Repatha® (Evolocumab) Available in the US at a 60 Percent Reduced List Price," news release, October 24, 2018, https://www.amgen.com/media/news-releases/2018/10/amgen-makes-repatha-evolocumab-available-in-the-us-at-a-60-percent-reduced-list-price/.

28. Amgen Corporation, "Amgen Makes Repatha® (Evolocumab) Available in the US at a 60 Percent Reduced List Price."

29. A. Gallo, "A Refresher on Marketing Myopia," *Harvard Business Review*, August 22, 2016, https://hbr.org/2016/08/a-refresher-on-marketing-myopia.

30. Orica Corporation, "Optimising Drill and Blast Operations with the Next Generation BlastIQ™ Digital Platform," news release, October 1, 2018, https://www.orica.com/news---media/optimising-drill-and-blast-operations-with-the-next-generation-blastiq-digital-platform.

31. Orica Corporation, "Optimising Drill and Blast Operations with the Next Generation BlastIQ™ Digital Platform."

32. P. Moore, "Orica's New BlastIQ™ Features Take Blast Digitalisation to the Next Level," *International Mining*, April 11, 2019, https://im-mining.com/2019/04/11/oricas-new-blastiq-features-take-blast-digitalisation-next-level/.

33. Orica Corporation, "Optimising Drill and Blast Operations with the Next Generation BlastIQ™ Digital Platform."

7 Committing to Outcomes

1. J. Pepitone, "America's Biggest Ripoffs," *CNN Money*, February 2, 2010, https://money.cnn.com/galleries/2010/news/1001/gallery.americas_biggest_ripoffs/6.html.

2. B. Popkin, "College Textbook Prices Have Risen 1,041 Percent Since 1977," *NBC News*, August 6, 2015, https://www.nbcnews.com/feature/freshman-year/college-textbook-prices-have-risen-812-percent-1978-n399926.

3. Pearson Company, "Pearson Launches the Learning Curve," news release, November 27, 2012, https://www.pearson.com/corporate/news/media/news-announcements/2012/11/pearson-launches-the-learning-curve.html#.

4. M. Barber, K. Donnelly, and S. Rizvi, "An Avalanche Is Coming: Higher Education and the Revolution Ahead," *IPPR* (March 2013): 14, 55.

5. Pearson 20-F filing with the US Securities and Exchange Commission, from the fiscal year ended Dec. 31, 2013, https://www.sec.gov/Archives/edgar/data/938323/000119312514117876/d693538d20f.htm.

6. Pearson promotional video, https://vimeo.com/138590287, accessed April 13, 2020.

7. Pearson, *MyLab Math for Developmental Math: Efficacy Research Report* (April 3, 2018), 13.

8. Pearson, MyLab: Math, https://www.pearsonmylabandmastering.com/northamerica/mymathlab/, accessed April 13, 2020.

9. "Seen at 11: Fitness Tracker Users Turn to Creative Hacks to Increase Step Counts," CBSN New York, August 22, 2016, https://newyork.cbslocal.com/2016/08/22/fitness-tracker-hacks/; J. Brown, "Fitness Fraud? How to Fool Your Fitness Tracker," Fox19Now.com, February 25, 2017, https://www.fox19.com/story/34601621/fitness-fraud-how-to-fool-your-fitness-tracker/.

10. J. Charniga, "Debt-Saddled Buyers Lean on Mom, Dad," *Automotive News*, March 10, 2019, https://www.autonews.com/sales/debt-saddled-buyers-lean-mom-dad.

8 Breaking the Quality Paradox

1. M. Mintz, "Still Hard to Swallow," *Washington Post*, February 11, 2001, https://www.washingtonpost.com/archive/opinions/2001/02/11/still-hard-to-swallow/013c5cb9-220e-451a-a327-852648342524/?utm_term=.d8b9e8dc0a30.

2. J. Hancock, "Senate Inquiry on Drug Prices Echoes Landmark Hearings Held 60 Years Ago," *Shots: Health News from NPR*, February 22, 2019, https://www.npr.org/sections/health-shots/2019/02/22/696967037/senate-inquiry-on-drug-prices-echoes-landmark-hearings-held-60-years-ago.

3. L. A. Caldwell, "In Senate Testimony, Pharma Executive Admits Drug Prices Hit Poor the Hardest," *NBC News*, February 26, 2019, https://www.nbcnews.com/politics/congress/senate-testimony-pharma-executive-admits-drug-prices-hit-poor-hardest-n976346.

4. A. Pollack, "Mylan Raised Epipen's Price before the Expected Arrival of a Generic," *New York Times*, August 25, 2016, https://www.nytimes.com/2016/08/25/business/mylan-raised-epipens-price-before-the-expected-arrival-of-a-generic.html.

5. J. Rockoff, "Mylan Reacts to Epipen Backlash," *Wall Street Journal*, August 26, 2016, https://www.wsj.com/articles/mylans-epipen-price-increases-highlight-its-grip-on-the-market-1472154769?mod=article_inline.

6. Rockoff, "Mylan Reacts to Epipen Backlash."

7. D. O'Neill and D. Scheinker, "Wasted Health Spending: Who's Picking Up the Tab?," *Health Affairs* (blog), May 31, 2018, https://www.healthaffairs.org/do/10.1377/hblog20180530.245587/full/.

8. O'Neill and Scheinker, "Wasted Health Spending."; National Academies of Sciences, Engineering and Medicine, "Transformation of Health System Needed to Improve Care and Reduce Costs," news release, September 6, 2012, http://www8.nationalacademies.org/onpinews/newsitem.aspx?RecordID=13444.

9. D. Roland, "At $2 Million, New Novartis Drug Is Priciest Ever," *Wall Street Journal*, May 24, 2019, https://www.wsj.com/articles/at-2-million-new-novartis-drug-is-priciest-ever-11558731506?mod=article_inline.

10. B. George, "Mylan CEO's Testimony Was a Huge Blow to the Entire Pharma Industry," *Biotech and Pharmaceuticals*, CNBC, September 27, 2016, https://www.cnbc.com/2016/09/27/mylan-ceos-testimony-was-a-huge-blow-to-the-entire-pharma-industry-commentary.html.

11. J. Walker, "Biotech Proposes Paying for Pricey Drugs by Installment," *Wall Street Journal*, January 8, 2019, https://www.wsj.com/articles/biotech-proposes-paying-for-pricey-drugs-by-installment-11546952520?mod=article_inline.

12. Walker, "Biotech Proposes Paying for Pricey Drugs by Installment."

13. J. Choi, G. W. Hecht, and W. B. Tayler, "Lost in Translation: The Effects of Incentive Compensation on Strategy Surrogation," *Accounting Review* 87, no. 4 (July 2012): 1135–1163.

14. M. Levine, "Fake Accounts Still Haunt Wells Fargo," *Bloomberg Opinion*, October 23, 2018, https://www.bloomberg.com/opinion/articles/2018-10-23/fake -accounts-still-haunt-wells-fargo.

15. M. Harris and B. Tayler, "Don't Let Metrics Undermine Your Business," *Harvard Business Review*, September–October 2019, https://hbr.org/2019/09/ dont-let-metrics-undermine-your-business; emphasis in original.

16. J. M. Izaret, D. Matthews, and M. Lubkeman, "Aligning Economic Incentives to Eradicate Diseases," BCG, January 7, 2019, https://www.bcg.com/publications/ 2019/aligning-economic-incentives-to-eradicate-diseases.aspx.

17. Izaret, Matthews, and Lubkeman, "Aligning Economic Incentives to Eradicate Diseases."

18. BCG, "An Innovative Approach to Pricing Drugs Can Accelerate the Eradication of Diseases," news release, January 14, 2019, https://www.bcg.com/ d/press/14january2019-an-innovative-approach-to-pricing-drugs-accelerate -eradication-disease-211735.

19. Izaret, Matthews, and Lubkeman, "Aligning Economic Incentives to Eradicate Diseases."

20. S. Kurczy, "Calculating the Benefits of Drugs," *Ideas and Insights*, Columbia Business School, February 12, 2019, https://www8.gsb.columbia.edu/articles/ ideas-work/calculating-benefits-drugs.

21. M. Nisen, "Why Big Pharma's Case to Congress Comes Up Short," *Bloomberg*, February 26, 2019, https://www.bloomberg.com/opinion/articles/2019-02-26/ drug-ceo-senate-testimony--why-big-pharma-s-case-comes-up-short-.

22. E. Barrette and K. Kennedy, "The Price-Quality Paradox in Healthcare," data brief, *IssueLab*, March 31, 2016, https://www.issuelab.org/resource/the-price -quality-paradox-in-healthcare.html.

9 Getting Up Close and Personal

1. "DNA Discount Advertisement for 'AeroMexico Airlines,'" *YouTube*, May 25, 2018, https://www.youtube.com/watch?v=2sCeMTB5P6U&feature=youtube, accessed April 13, 2020.

2. L. Marcus, "AeroMexico's New 'DNA Discount' Ad Goes Viral," CNN, January 18, 2019, https://www.cnn.com/travel/article/aeromexico-dna-discount-travel-ad -video/index.html.

3. B. Keveney, "Aeromexico Ad Campaign Trolls Anti-Mexican Sentiment in US with DNA Discounts," *USA Today*, January 19, 2019, https://www.usatoday.com/story/travel/2019/01/17/aeromexico-tweaks-u-s-dna-discount-campaign/2609363002/; S. Van Sant, "How the Partial Government Shutdown Could Affect You," NPR *Morning Edition*, December 23, 2018, https://www.npr.org/2018/12/23/679652640/how-the-partial-government-shutdown-could-impact-you.

4. A. Garcia, "Is Aeroméxico Offering 'DNA Discounts' for People Traveling to Mexico?," Snopes, January 21, 2019, https://www.snopes.com/fact-check/aeromexico-dna-discount/.

5. D. Griner, "A Viral Hit, 7 Months Delayed: Aeroméxico's DNA Ad Shows the Unpredictability of Organic Video," *AdWeek*, January 22, 2019, https://www.adweek.com/creativity/a-viral-hit-7-months-delayed-aeromexicos-dna-ad-shows-the-unpredictability-of-organic-video/.

6. The quote makes a valid point, even though Sutton denies he ever said it: https://www.snopes.com/fact-check/willie-sutton/, accessed April 13, 2020.

7. P. Ausick, "Billions of Records Exposed: 2019 On Track to Be Worst Year Ever for Data Breaches," *USA Today*, August 19, 2019, https://www.usatoday.com/story/money/2019/08/18/2019-on-track-to-become-worst-year-ever-for-data-breaches/39963021/.

8. J. Davis, "The 10 Biggest Healthcare Data Breaches of 2019, So Far," Health IT Security, July 23, 2019, https://healthitsecurity.com/news/the-10-biggest-healthcare-data-breaches-of-2019-so-far.

9. FBI, "Bank Crime Statistics 2018," *Documents*, https://www.fbi.gov/file-repository/bank-crime-statistics-2018.pdf/view, accessed April 13, 2020.

10. Accenture, "U.S. Consumers Turn Off Personal Data Tap as Companies Struggle to Deliver the Experiences They Crave," news release, December 5, 2017, https://newsroom.accenture.com/news/us-consumers-turn-off-personal-data-tap-as-companies-struggle-to-deliver-the-experiences-they-crave-accenture-study-finds.htm.

11. S. Maheshwari, "How Smart TVs in Millions of U.S. Homes Track More Than What's on Tonight," *New York Times*, July 5, 2018, https://www.nytimes.com/2018/07/05/business/media/tv-viewer-tracking.html.

12. Salesforce, *"The State of the Connected Consumer 2nd Edition,"* 2018, 5, https://www.salesforce.com/ap/form/conf/state-of-the-connected-customer-2nd-edition/, accessed July 6, 2016.

13. A. Ghose, *TAP: Unlocking the Mobile Economy* (Cambridge, MA: MIT Press, 2016).

14. Syncron, https://www.syncron.com/company/, accessed April 13, 2020.

15. L. Bell, "Uptime: The Most Disruptive Change in Pricing History," Syncron, May 4, 2018, https://www.syncron.com/uptime-the-most-disruptive-change-in -pricing-history/.

16. C. Weiss, S. Gaenzle, and M. Römer, "How Automakers Can Survive the Self-Driving Era," A.T. Kearney, 2015, 4, https://www.kearney.com/documents/ 20152/434078/How%2BAutomakers%2BCan%2BSurvive%2Bthe%2BSelf -Driving%2BEra%2B%25282%2529.pdf/3025b1a0-4d71-e24d-51e0-2cc1f290447c ?t=1493941955625.

17. S. Corwin, N. Jameson, D. Pankratz, and P. Willigmann, "The Future of Mobility: What's Next?," *Insights*, Deloitte, September 14, 2016, https://www2 .deloitte.com/us/en/insights/focus/future-of-mobility/roadmap-for-future-of -urban-mobility.html?icid=dcom_promo_standard|us;en.

18. Weiss, Gaenzle, and Römer, "How Automakers Can Survive the Self-Driving Era," 19.

19. D. Langkamp, J. Schürmann, T. Schollmeyer, R. Kilian, A. Petzke, J. Pineda, and Jean-Manuel Izaret, "How the Internet of Things Will Change the Pricing of Things," BCG, December 7, 2017, https://www.bcg.com/publications/2017/ how-internet-of-things-change-pricing-of-things.aspx.

20. Philips, "Circular Lighting at Schiphol Airport," global professional lighting website, https://www.lighting.philips.com/main/cases/cases/airports/ schiphol-airport, accessed April 13, 2020.

10   Partnering with Customers

1. M. Roy and M. Liersch, "I Am a Better Driver Than You Think: Examining Self-Enhancement for Driving Ability," *Journal of Applied Social Psychology* 43 (2013), https://www.ncbi.nlm.nih.gov/pmc/articles/PMC3835346/ - R46; J. Kruger and D. Dunning, "Unskilled and Unaware of It: How Difficulties in Recognizing One's Own Incompetence Lead to Inflated Self-Assessments," *Journal of Personality and Social Psychology* 77 (1999): 1121–1134, https://www.ncbi.nlm.nih.gov/ pubmed/10626367.

2. A. F. Williams, "Views of U.S. Drivers about Driving Safety," *Journal of Safety Research* 34 (2003): 491–494, https://www.sciencedirect.com/science/article/pii/ S0022437503000690.

3. "When It Comes to Driving, Most People Think Their Skills Are above Average," Association for Psychological Science, August 28, 2014, https://www .psychologicalscience.org/news/motr/when-it-comes-to-driving-most-people -think-their-skills-are-above-average.html.

4. Progressive, "Snapshot FAQ," https://www.progressive.com/auto/discounts/snapshot/snapshot-common-questions/, accessed June 6, 2019.

5. Progressive, "Snapshot FAQ."

6. Progressive, "Snapshot FAQ."

7. Progressive, "Get Snapshot from Progressive," signup and discounts, https://www.progressive.com/auto/discounts/snapshot/, accessed June 6, 2019.

8. B. Yerak, "Progressive Using Snapshot to Add Surcharge for Aggressive Drivers," *Chicago Tribune*, March 27, 2015, https://www.chicagotribune.com/business/ct-progressive-aggressive-surcharge-0330-biz-20150327-story.html.

9. Progressive, "Get Snapshot from Progressive."

10. Progressive, "New Progressive Data Shows Putting the Phone Down Correlates to Lower Insurance Claims," news release, January 17, 2019, https://progressive.mediaroom.com/2019-01-17-new-progressive-data-shows-putting-the-phone-down-correlates-to-lower-insurance-claims.

11. Progressive, "Snapshot FAQ."

12. Progressive, "Check Your Latest Status," Snapshot account log-in, https://www.progressive.com/snapshot/tdx/Account/Logon, accessed June 6, 2019.

13. U.S. Food and Drug Administration, "Best Practices for Tablet Splitting," August 23, 2013, https://www.fda.gov/Drugs/ResourcesForYou/Consumers/BuyingUsingMedicineSafely/EnsuringSafeUseofMedicine/ucm184666.htm.

14. K. Arrow, "Uncertainty and the Welfare Economics of Medical Care," *American Economic Review* 53 (December 1963), https://www.who.int/bulletin/volumes/82/2/PHCBP.pdf.

15. A. Schechter, "There Is Regulatory Capture, but It Is by No Means Complete," *ProMarket* (blog), March 15, 2016, https://promarket.org/there-is-regulatory-capture-but-it-is-by-no-means-complete/.

16. D. Lancefield and C. Gagliardi, "What You Need to Know before You Sign a Payment-by-Results Contract," *Harvard Business Review*, September 5, 2016, https://hbr.org/2016/09/what-to-know-before-you-sign-a-payment-by-results-contract.

17. D. Frydlinger, O. Hart, and K. Vitasek, "A New Approach to Contracts," *Harvard Business Review*, September–October 2019, https://hbr.org/2019/09/a-new-approach-to-contracts.

18. P. Clist and S. Dercon, "12 Principles for Payment by Results (PbR) In International Development," Department for International Development, June 30, 2014, https://assets.publishing.service.gov.uk/media/57a089d2e5274a27b20002a5/clist-dercon-PbR.pdf.

19. W. Grad, J. Grudnowski, E. Spaulding, and S. D. Burton, "The Measurement Advantage," Bain & Company, March 26, 2019, https://www.bain.com/insights/the-measurement-advantage/.

20. Lenntech, "Water and Process Solutions for the Brewing Industry," data sheet, https://www.lenntech.com/Data-sheets/GE-solutions-00GEA19623-Brewery-L .pdf.

21. P. Klement, "Gamification: Mehr als Belohnungen und Badges," *Techtag*, September 10, 2018, https://www.techtag.de/unkategorisiert/gamification-mehr -als-belohnungen-und-badges/.

22. R. M. Thaler and C. Sunstein, *Nudge: Improving Decisions about Health, Wealth, and Happiness* (New York: Penguin Group, 2008), 6.

23. Mitsubishi Motors, "Mitsubishi Motors and LexisNexis Risk Solutions Provide Driver Feedback Notifications and Usage-Based Insurance (UBI) Options Through Road Assist+ App," news release, June 5, 2018, https://media .mitsubishicars.com/releases/mitsubishi-motors-and-lexisnexis-risk-solutions -provide-driver-feedback-notifications-and-usage-based-insurance-ubi-options -through-road-assist-app?page=10.

24. C. Dawson, "Mitsubishi Bets People Will Reveal Their Driving Habits to Insurers—For a Freebie," *Wall Street Journal*, July 6, 2018, https://www.wsj.com/ articles/mitsubishi-bets-people-will-reveal-their-driving-habits-to-insurersfor-a -freebie-1530853415.

25. Dawson, "Mitsubishi Bets People Will Reveal Their Driving Habits to Insurers—For a Freebie."

26. Starbucks "Stars" reward program, https://www.starbucks.com/rewards/ #starcode, accessed April 13, 2020.

27. Red Sox Rewards, http://redsox.info-mlb.com/rewards/#rewards, accessed April 13, 2020.

28. N. Meyersohn, "Who Knew, Tide Does Dry Cleaning. Now It's Expanding the Business," *CNN Business*, February 20, 2019, https://www.cnn.com/2019/02/ 20/business/tide-cleaners-laundry-service/index.html.

29. S. Terlep, "P&G Starts Online Subscription Service for Tide Pods," *Wall Street Journal*, July 20, 2016, https://www.wsj.com/articles/p-g-starts-online-subscription -service-for-tide-pods-1468941036.

30. Tide Cleaners, https://www.tidecleaners.com/, accessed April 13, 2020.

31. Meyersohn, "Who Knew, Tide Does Dry Cleaning."

32. S. Bhattacharyya, "P&G Is Testing a Tide-Branded Laundry Service," *Digiday*, February 20, 2019, https://digiday.com/retail/pg-testing-tide-branded -laundry-service/.

11 Making Your Move

1. A. Diallo, "Adobe's Subscription-Only CC Release Carries Obvious Upside But Big Risk," *Forbes*, June 17, 2013, https://www.forbes.com/sites/amadoudiallo/

2013/06/17/adobe-cc-subscription-release-big-upside-and-risk/#122d226a19c6; Adobe 10-K report for the fiscal year ended November 30, 2012.

2. "Adobe's Shift to the Cloud: Is This the Start of a Trend?," *Knowledge@ Wharton*, May 8, 2013, http://knowledge.wharton.upenn.edu/article/adobes -shift-to-the-cloud-is-this-the-start-of-a-trend/.

3. D. Wadhwani, as quoted in S. Shankland, "Dislike Adobe's Creative Cloud Subscriptions? Tough Beans," cnet, May 28, 2013, https://www.cnet.com/news/ dislike-adobes-creative-cloud-subscriptions-tough-beans/.

4. Adobe 10-K report for the fiscal year ended November 30, 2012.

5. Some dispute whether Gillette invented the razor and blades business model or defaulted to it. For an interesting perspective, see https://hbr.org/2010/09/ gillettes-strange-history-with.

6. R. Trenholm, "Your Printer Orders Ink for You with New HP Instant Ink Service," cnet, May 28, 2014, https://www.cnet.com/news/hp-instant-ink/.

7. T. Poletti, "How HP Is Trying to Resuscitate a Declining Business," *Market-Watch*, July 3, 2017, https://www.marketwatch.com/story/how-hp-is-trying-to -resuscitate-a-dying-business-2017-06-30.

8. T. Hoffman, "How to Save Money with HP Instant Ink and Other Low-Cost Printer Ink Program," *PCMag*, December 12, 2017, https://www.pcmag.com/ article/357838/how-to-save-money-with-hp-instant-ink-and-other-low-cost -pri.

9. HP Inc. (HPQ), Q1 2019 Earnings Conference Call Transcript, *The Motley Fool*, February 27, 2019, https://www.fool.com/earnings/call-transcripts/2019/ 02/27/hp-inc-hpq-q1-2019-earnings-conference-call-transc.aspx.

10. "Newton's Laws of Motion," National Aeronautics and Space Administration, https://www.grc.nasa.gov/www/k-12/airplane/newton.html, accessed April 13, 2020.

11. J. Bezos, as quoted in J. Del Rey, "This Is the Jeff Bezos Playbook for Preventing Amazon's Demise," *Vox*, April 12, 2017, https://www.vox.com/2017/4/12/ 15274220/jeff-bezos-amazon-shareholders-letter-day-2-disagree-and-commit.

Index